Aspects of modern sociology

THE SOCIAL STRUCTURE
OF MODERN BRITAIN

GENERAL EDITORS

John Barron Mays
Eleanor Rathbone Professor of Sociology
University of Liverpool

Maurice Craft
Professor of Education
University of Nottingham

ASPECTS OF MODERN SOCIOLOGY

General Editors

John Barron Mays Professor of Sociology, University of Liverpool
Maurice Craft Professor of Education, University of Nottingham

This Longman library of texts in modern sociology consists of three Series, and includes the following titles:

THE SOCIAL STRUCTURE OF MODERN BRITAIN

The family
Mary Farmer
University of Liverpool

The political structure
Grace Jones
King Alfred's College,
Winchester

Population
Prof. R. K. Kelsall
University of Sheffield

Education
Ronald King
University of Exeter

The welfare state
Prof. David Marsh
University of Nottingham

Crime and its treatment
Prof. John Barron Mays
University of Liverpool

Patterns of urban life
Prof. R. E. Pahl
University of Kent

The working class
Kenneth Roberts
University of Liverpool

The middle class
John Raynor
The Open University
and
Roger King
Huddersfield Polytechnic

Leisure
Kenneth Roberts
University of Liverpool

Adolescence
Cyril Smith
Social Science Research Council

The mass media
Peter Golding
University of Leicester

The legal structure
Michael Freeman
University of London

Rural life
Gwyn Jones
University of Reading

Religious institutions
Joan Brothers
University of London

Mental illness
Bernard Ineichen
University of Bristol

The economic structure
Prof. Cedric Sandford
University of Bath

SOCIAL PROCESSES

Bureaucracy
Dennis Warwick
University of Leeds

Social control
C. Ken Watkins
University of Leeds

Communication
Prof. Denis McQuail
University of Amsterdam

Stratification
Prof. R. K. Kelsall
University of Sheffield
and
H. Kelsall

Industrialism
Barry Turner
University of Exeter

Social change
Anthony Smith
University of Reading

Socialisation
Graham White
University of Liverpool

Social conflict
Prof. John Rex
University of Warwick

Forthcoming titles will include:

Migration
Prof. J. A. Jackson
University of Dublin

SOCIAL RESEARCH

The limitations of social research
Prof. M. D. Shipman
University of Warwick

Social research design
Prof. E. Krausz
University of Newcastle
and
S. H. Miller
City University

Sources of official data
Kathleen Pickett
University of Liverpool

History of social research methods
Gary Easthope
University of East Anglia

Deciphering data
Jonathan Silvey
University of Nottingham

The philosophy of social research
John Hughes
University of Lancaster

Data collection in context
Stephen Ackroyd
and
John Hughes
University of Lancaster

The Economic Structure

CEDRIC SANDFORD M.A.(Econ)
Professor of Political Economy
Director, Centre for Fiscal Studies
University of Bath

Longman London and New York

Longman Group Limited
Longman House
Burnt Mill, Harlow, Essex, UK

*Published in the United States of America
by Longman Inc., New York*

© Longman Group Limited 1982

All rights reserved. No part of this publication may be
reproduced, stored in a retrieval system, or transmitted
in any form or by any means, electronic, mechanical,
photocopying, recording, or otherwise, without the
prior permission of the Copyright owner.

First published 1982

British Library Cataloguing in Publication Data

Sandford, Cedric
 The economic structure. - (Aspects of modern sociology. Social
 structure of modern Britain)
 1. Great Britain - Economic conditions - 1945-
 I. Title. II. Series.
 330.941'0858 HC256.6

 ISBN 0-582-29544-0

Library of Congress Cataloging in Publication Data

Sandford, Cedric
 The economic structure. - (Aspects of modern sociology. Social
 structure of modern Britain)
 Bibliography: p.
 Includes index
 1. Great Britain - Economic conditions - 1945-
 I. Title. II. Series.
 HC256.6.S313 330.941'0858 81-20751

 ISBN 0-882-29544-0 AACR2

Printed in Singapore by
Selector Printing Co Pte Ltd.

CONTENTS

Editors' Preface viii
Foreword ix
By the same author xi
Acknowledgements xii

1 The science of economics 1
2 The national economy 13
3 Solving the economic problem: the market system 26
4 Solving the economic problem: the centrally planned economy 37
5 The mixed economy: the United Kingdom 45
6 The private sector in the United Kingdom 56
7 The public sector in the United Kingdom 75
8 The scope and limits of government policy 90

Notes 105
References and further reading 107
Index 109

EDITORS' PREFACE

This series has been designed to meet the needs of students following a variety of academic and professional courses in universities, polytechnics, colleges of higher education, colleges of education, and colleges of further education. Although principally of interest to social scientists, the series does not attempt a comprehensive treatment of the whole field of sociology, but concentrates on the social structure of modern Britain which forms a central feature of most such tertiary level courses in this country. Its purpose is to offer an analysis of our contemporary society through the study of basic demographic, ideological and structural features, and the examination of such major social institutions as the family, education, the economic and political structure and religion. The aim has been to produce a series of introductory texts which will in combination form the basis for a sustained course of study, but each volume has been designed as a single whole and can be read in its own right.

We hope that the topics covered in the series will prove attractive to a wide reading public and that, in addition to students, others who wish to know more than is readily available about the nature and structure of their own society will find them of interest.

John Barron Mays
Maurice Craft

FOREWORD

This book attempts to portray the framework within which the economic activities of a society are conducted. It is written for students with no previous knowledge of economics, so economic terms are explained as they are introduced. It is intended primarily for social science students of university level who are not specialising in economics; but the author hopes it may prove of value to some 'A' level students, students in further education and to the elusive general reader. Inevitably this portrayal requires some description of institutions and organisations; and equally inevitably, because of the size of the book, these descriptions are restricted to one country, the United Kingdom – though where possible brief points of international comparison are included to widen the perspective. The emphasis of the book, however, is less on description than on analysis. It seeks to explain how the institutions have arisen, what functions they perform, what deficiencies they suffer from and, where possible, how they may be improved.

With such a wide compass explanations must necessarily be compressed and therefore simplified. This carries dangers of (unintentional) misrepresentation. However, the book should properly be regarded as no more than an introduction to a huge subject area, as much concerned with raising questions as with answering them.

In any book which covers a broad aspect of the life of a society, in explanations of the past and still more in prescriptions for the future, value judgements intrude. The best an author can do is to make the reader aware of alternative viewpoints and to be open about his own philosophy and predilections. The author might describe himself as a liberal-social-democrat. He would contend that Marxist analysis has furnished important insights about the nature of society but is very far from the whole truth. He is a supporter of the 'mixed economy' but is less than happy about the nature of the mix. As a liberal he considers that the nature and virtues of the free market are inadequately appreciated and the wider application of pricing principles is desirable to promote efficiency and well-being. As a socialist (using that word in the broad sense that 'Socialism is about equality', not in the narrow sense of concern with the ownership of the means of production, distribution and exchange) he wishes to see a reduction of the inequalities of income and wealth in society, brought about by evolution not revolution. As a democrat he would like to see a political system which would better reflect the people's views and the introduction of additional elements of democracy into economic life. These are rather vague generalisations; the contents of the book furnish some of the concrete applications.

If we are to give an account of the 'economic structure' we must start by saying something about the meaning and nature of economics and how economic activity is conducted and measured. That is the theme of the first two chapters. Chapters 3 and 4 are concerned with alternative extreme solutions to the economic problem of what to produce, how to produce it, and who gets the product: the *laissez-faire* market system on the one hand and the centrally planned economy on the other. In Chapter 5 we look at the reality of the 'mixed economy' of the UK with its particular balance of public and private sectors, examining the growth of the public sector and the characteristics of private and public goods. In the following chapters we examine in more detail the private and public sectors respectively. In the final chapter we survey the scope and limitations of government policy measures in the mixed economy examining, in particular, the distribution of income and wealth and the problems of unemployment and inflation.

Throughout the book the masculine 'he' or 'his' has been used for illustrative examples when the feminine would have served equally well. Readers with understandable susceptibilities on this issue are invited to substitute their own 'she' and 'hers'. To have written 'he or she' and 'his or hers' every time would have been tedious; to have alternated the masculine and the feminine would have been pedantic and confusing; so the author has fallen back on convention, with apologies to all female readers.

Cedric Sandford

BY THE SAME AUTHOR

Realistic Tax Reform (1971)
Taxing Personal Wealth (1971)
An Annual Wealth Tax (joint author) (1975)
National Economic Planning (1976)
The Economics of Public Finance (1978)
Taxation and Social Policy (joint ed.) (1981)
Costs and Benefits of VAT (joint ed.) (1981)

ACKNOWLEDGEMENTS

The considerable indebtedness of the author to other writers is acknowledged by way of references. Beyond that, the author is aware of an obligation, which he readily acknowledges, to many colleagues and even more students - an obligation nonetheless real for being imprecisely definable. Finally, his thanks are due, and very warmly expressed, to Sue Powell and Barbara Hall for typing the manuscript with exemplary speed and accuracy.

1
THE SCIENCE OF ECONOMICS

The meaning of economics

'The study of mankind in the ordinary business of life' was how Alfred Marshall, a famous late nineteenth- and early twentieth-century economist, defined economics. This definition is a useful starting point, indicating the pervasiveness of economic activity; but it is rather too general and vague; moreover it is a descriptive type of definition which fails to get to the essence of the subject.

In seeking an analytical definition we might do worse than to say, at the risk of tautology, that economics is the science of economising; that it is concerned with using scarce resources to best advantage. More generally, economics is the science that studies how man uses the resources at his disposal to satisfy his wants.

Each element in this definition requires elaboration: we need to say something more about economics as a science; about the meaning and characteristics of scarce resources; and about the wants or ends.

Economics as a science

When we describe economics as a science we are primarily saying something about methods and objectives. A science is a systematised branch of knowledge which develops by the two methods of deduction and induction; on the one hand logical thinking in one's armchair; on the other hand, getting out and looking at the real world. By deduction the implications of definitions and assumptions are worked out and theories are formulated and developed. But the essence of science is the appeal to fact and the counterpart of deduction is observation, recording and experimentation. Such empirical work tests the validity of theories, leads to their refinement and may suggest new theoretical possibilities.

Science is very much concerned with measurement. Economics, like other sciences, seeks to evolve laws or generalisations about what will happen under certain circumstances, like the law of gravity in physics or laws of supply and demand in economics. Often, at the present stage of development of economics, these laws are qualitative only, indicating the direction of change but not its extent. But the ideal is to make them quantitative; for example, we would like to say not only that a restriction in the supply of a particular commodity would cause its price to rise, but precisely how large a price rise would be associated with a specified supply reduction.

By developing laws and appreciating the inter-connectedness of the economy the economist, as scientist, seeks to predict, to indicate what will happen under certain circumstances. He seeks to say, for example, what will happen to output, employment and prices in the economy over the next twelve months. The purpose of the prediction is control; if the prediction is unpalatable then we can attempt to change the outcome by appropriate policy measures.

Because economics is a science concerned with human behaviour it comes under the heading of social science. Social sciences suffer from distinct limitations, as compared with natural sciences, arising from the nature of their subject matter. The social scientist is seeking to predict the behaviour of heterogeneous people while the natural scientist deals with homogeneous matter. This difference does not mean that the social scientist cannot predict. Prediction is concerned with people in the mass, where the idiosyncrasies tend to cancel out or not to matter. Thus the generalisation that 'when average real incomes of a particular community rise more meat will be bought' is not gainsaid because some citizens are vegetarians – but it does make prediction more hazardous. Again, it is rarely possible in the social sciences to conduct experiments like the laboratory experiments of the physicist. We can rarely put people in a carefully controlled environment to see how they behave in response to some isolated stimulus, while we hold other relevant factors constant. Psychologists sometimes attempt this kind of experiment with individuals, but because the subjects of the experiment are aware that the situation is artificial their reactions may be abnormal.

It is probably also true that the 'other things' to be held constant, if we are to isolate the effects of one particular factor, are more numerous in the social sciences. The simple demand situation which we earlier postulated, of an increase in average incomes generating a rise in demand for meat, illustrates the problem. Incomes do not usually rise instantaneously but gradually over a period. During that period a vast number of other changes may take place. The prices of fish and eggs, which are to some extent substitutes for meat, might fall. Or, under the influence of ecologists or the RSPCA, more people might become vegetarians. Or people might become more religious and eat fish on Fridays. It might be discovered that fish have more nutrients than meat. And so on. In practice we can rarely identify all the influences of a comparatively simple issue like that, let alone control them. How much more difficult it is to make an accurate prediction about the extent of a slump in the economy, or when an upturn of activity will occur.

Another feature of the social sciences is that their subject matter, people, learn from experience; as a result they may not always behave in the same way in the same circumstances. For example, one way for a government to deal with a temporary slump in an industry would be to remove any sales tax from the product so that the price fell and more was purchased. In boom times the tax could be reimposed. For a while such a policy might work to even out activity and employment in that industry but, because people learn from experience, the time might come when people might react to a slight dip in the industry's sales by holding back on purchases in anticipation of a tax reduction. Then their action would precipitate a slump in the industry which might not otherwise have happened. Because people learn from experience a policy intended to stabilise output and employment would have become destabilising.

Another related problem is that of expectations. If enough people think something is going to happen it will happen. If investors think that interest rates are going to fall then they will move into the market to buy irredeemable or long-dated government securities but this raises the prices of the government securities.[1] A government bond which was selling for £100 and on which the government promised to pay £10 per year in interest might rise in price to £110. The rate of interest which was previously 10 per cent (£10 in £100) has now fallen to just over 9 per cent (£10 in £110). If the government wants to borrow new money it only needs to offer 9 per cent on any £100 of new stock it puts onto the market.

Or take the case of unemployment. If managers think there is going to be a slump in demand they will cut back orders for new equipment. If workers think that unemployment will increase, they may reduce their purchases of goods and services in order to save against a 'rainy day'. But the fall in demand for equipment and for goods and services will lead to unemployment among the workers making the products. Thus the expectation of slump and unemployment has created a slump and unemployment.

Expectations are particularly important for inflation. If people expect prices to rise rapidly then they will build this expectation into their economic bargains: workers will demand higher wages to compensate for the expected price increase; lenders will require higher interest for the same reason; engineering and construction firms seeking contracts will write a contingency allowance into their tender prices against anticipated cost increases; and so on – with the effect that the expectation of rapidly rising prices generates rapidly rising prices. This is one reason why inflation is so hard to get rid of once it has taken hold.

All this adds to the complications of the social sciences, but also to their interest and fascination. Nor is the social scientist naked in the face of these difficulties. If he cannot conduct controlled experiments with a particular group of people he may be able to compare two 'matched' samples of people. For example, in studying the economics of education we should like to know what is the effect of a university degree on people's lifetime earnings. Earnings are affected by intelligence and upbringing, to mention the most obvious factors, as well as education. We cannot control the background, IQ and education of a particular group of people to see what difference extra education makes, but we can compare two groups of people of similar background and IQ, but with the difference that one group has received three years of higher education and the other has not. Then a comparison of the average lifetime earnings of the people in these groups will give us the 'education differential'.

Because of our general inability to conduct controlled experiments in the social sciences we tend to rely heavily on statistics – in two senses. One meaning is just plain numbers. We collect data and compare it over time (a time series) to see if there is any pattern or (as in measuring the differential earnings from different amounts of education) we compare data for different situations at the same time (cross section comparisons). Statistics in this sense of quantitative measures are indispensable in the social sciences to provide a concise summary of a complex situation and as a factual basis for explanation, diagnosis and policy formulation; but they need to be used with great care if they are not to mislead.[2] In particular the social scientist nearly always has to use statistics which have been collected for administrative purposes or as a by-product of administration and rarely do they exactly match his requirements.

The second sense of the word 'statistics' is that of statistical techniques. These techniques vary from the simple, such as the arithmetical average or mean, to the sophisticated, such as regression analysis, which attempts to isolate the effects of each of a number of different influences in a situation.

Economics is the most developed of the social sciences. One branch of economics, econometrics, is wholly concerned with giving precise quantities to economic variables. But even economics, by comparison with physics, 'has hardly yet reached its seventeenth century' (Phelps Brown 1972). There remains within the social sciences, including economics, wide scope for the exercise of judgement. It is hardly surprising in the complex world of today that there are big differences in interpretation and in prognostication.

Scarce resources

Economics, then, is a social science. What distinguishes it from other social sciences is that it is concerned with the use of scarce resources to satisfy wants. What are these scarce resources?

If this question was put to 'the man on the Clapham omnibus' his most likely answer (assuming he was disposed to be civil) would be 'money'. Pressed a little he might modify that to 'money income'. Within its limitations this answer is perfectly valid. We each think of our money income as the constraint on satisfying our wants. If we had more money we could satisfy more wants.

We shall be saying a lot more about money later in the book; for the moment let us identify it as a highly convenient medium of exchange. Money does not satisfy our wants directly, but it is a convenient means by which we can obtain goods and services. However, there would in fact be no technical difficulty about providing all of us with more money income – the government could simply set the printing presses to work and send us all £100 per week through the post. But it wouldn't do us much good. As we all delightedly rushed to the shops to spend this income, we should find the goods disappearing from the shelves and long queues appearing at the cinemas. Before long prices would start to rise. The increased demand as a result of the extra money might lead to some increase in the total of goods supplied if there were previously unemployed workers and machines, but the main effect would be a rise in prices. We would have solved the scarcity of money only to find there was a scarcity of goods.

Are goods and services, then, the scarce resources to which the definition of economics refers? In an immediate sense this is so, but there is a more ultimate and important sense. Let us refer back to the previous examples. We said that, following the increase in money income, there might be some increase in the supply of goods to the shops if there were unemployed workers who could be brought into employment and unemployed machines which could be brought into use. Indeed more goods could be produced if workers in employment worked overtime and if machinery already in use was used more intensively. But capacity working would soon be reached. Here then is our answer: the ultimate scarce resources are the workers and the equipment necessary to produce the goods and services. We can apply this finding to the individual. The scarce resource which limits the money income of the individual wage earner is the amount of labour power he can supply and the value society puts on it. For the property owner it is the amount he can lend to the productive system and the value placed on that. Where income is derived from a social security benefit the relationship between income and the scarce resources is more indirect; but the size of social security benefits is limited by how many scarce resources the society is able and willing to see transferred to the social security beneficiary from the other people in the community.

Traditionally, economists have called the scarce resources the agents or factors of production and grouped them into three broad categories, Labour, Capital and Land. There are many more than three factors of production, for each of these categories includes a wide range of factors; but it is a convenient designation, for all the factors within each category have a common characteristic.

Thus Labour is the immediate human element in production. It includes the manual and non-manual; the unskilled and the skilled of many different sorts; and it includes judgement, flair and enterprise (which are sometimes given a separate group identity).

Capital consists of man-made equipment of all kinds – factories, machines, roads, hospitals, schools, houses and also stocks of semi-finished goods.

Land can be thought of as a rather special kind of equipment, one which is provided by nature. The term 'Land' is not confined to agricultural or site land, but extends to mineral resources of all kinds.

The dividing line between Land and Capital is somewhat blurred in practice. One famous economist, Ricardo, thought the distinction lay in the 'natural and indestructible' qualities of the soil. This was an unfortunate phrase. Once a farm has been cultivated it becomes impossible to distinguish how much of its fertility is due to nature and how much to the application of artificial fertilisers. As to indestructibility, we can call on many examples of how bad husbandry can destroy natural qualities, in the extreme case leading to dust bowls. The dividing line with mineral resources, coal for example, is hardly more clear-cut. Once extracted from the soil and (say) stocked in a factory awaiting use, we normally count raw materials as capital. The really important distinction between Land and other forms of equipment is that, broadly speaking, land is fixed in extent. It is possible to push back the sea here and there and to go on for a very long time finding new sources of non-renewable materials; but Land as we have defined it is subject to supply limits which do not apply to man-made capital – and that is a fact of fundamental importance.

There are a few more comments we need to make about the characteristics of scarce resources in general. Indeed, the concept of scarcity itself needs clarification. Something is not scarce simply because there is only a little of it. Its scarcity is always relative to the uses to which it can be put, or, as the economist would say, scarcity is relative to demand (which is desire backed by purchasing power). To borrow an example from Professor Robbins, there are more good eggs than bad, but only good eggs are scarce. That generalisation may require some minor modification at General Election times, but its validity is indisputable in normal times. Again, if a particular highly specialised manufacturing process, for which only a few machines existed, was made obsolete by new technology, the machines might be rendered valueless (or reduced to scrap value); there might only be a few of them, but that would not make them scarce.

A supremely important characteristic of the factors of production is that, within limits, they can be used in different ways. Take Labour; workers can make all sorts of different goods or provide a range of different services. Many factory workers have skills of manual dexterity which can be applied to a range of different products; and a clerk can perform clerical functions equally well in a variety of different industries. Of course, not all workers have the aptitude or skills to take on all jobs: but with some re-education and re-training a major realignment of the labour force could take place in a very short time.

Many items of capital can be used to produce different goods. Some equipment, like a blast furnace, produces an output which can then be used for a variety of different products. Buildings can be modified to provide flats or offices. Even fairly specific machinery, like looms, can be used to make a variety of different kinds of cloth. There is some machinery which is highly specific and cannot be adapted for other uses, but there is an ultimate flexibility with capital in that, as a machine wears out, it need not be replaced by an identical machine.

Even Land is capable of alternative uses. Land may be used for agriculture, building or recreational purposes. Agricultural land can be used to grow a wide variety of different human foodstuffs, animal foodstuffs, or raw materials for industry. Site land may be used for houses, flats, offices, hospitals, schools, aerodromes or army barracks. And so on. Whilst there are some constraints, an essential characteristic of the factors of production is that they can be used in a wide variety of different ways.

A final point: time itself is often thought of as being a scarce resource, through its scarcity putting constraints on what wants can be satisfied, and being eminently capable of alternative use.

Wants or ends

The scarce resources are used to satisfy wants or ends. But what do these imply?

Wants are many and varied. The ends we seek may be individual, family or household, like food, shelter or clothing; or they may be communal objectives such as more hospitals, schools and universities.

The wants or objectives may be phrased in general terms, like 'doubling the standard of living in twenty-five years', or in precise terms like 'building three hundred thousand houses in a year'. The wants may include the immaterial as well as the material, e.g. a person may wish to listen to live music concerts where the product vanishes in the moment of its creation. Not only are the wants multifarious, they are virtually unlimited. Few, if any, of us can honestly say that all our wants or ends are satisfied. Nor need our ends be selfish. We may seek more resources in order to use them for the benefit of others. But, whether selfish or altruistic, the fundamental characteristic of wants is that they are virtually unlimited.

The second point of importance is that it is not for the economist, as an economist, to pass judgement on the ends. A particular government may seek to increase the size and improve the quality of the national defence forces. The economist as a citizen may approve or deplore this policy, but as an economist his concern is limited to showing the resource implications of the policy and to demonstrating how to achieve the objective as economically as possible, i.e. with the least expenditure of scarce resources. In other words, economics is about means, not ends. This limitation does not make economics either ignoble or unimportant. If the economist can show how a particular objective can be achieved with fewer resources, then more are available to satisfy other ends. Also, in showing the resource implications of particular policies he may highlight policy inconsistencies which can then be remedied. These are important contributions to welfare.

To the view that economists as such are not concerned with ends, it may be objected that most policy measures on which economists are called on to advise require or imply some value judgements. For example, later in the book we shall be looking at the distribution of income and wealth in the community and suggesting ways in which inequality might be reduced. Yet the objective of reducing inequality of income and wealth is essentially based on a value judgement. However, in effect, if not always explicitly, when the economist looks at the implications of alternative ways of reducing inequality he is making propositions of the kind 'if the community wishes to reduce inequality in the distribution of income or wealth, then here are alternative ways of doing it with the following implications . . .'. Even so we must accept that the dividing line between ends and means is not always easy to draw and that, in restricting the range of the policy alternatives considered, the economist may implicitly, and perhaps unconsciously, be making a value judgement about the acceptability of the broad framework of the economic system within which he works.

Economics and choice

In our first paragraph we defined economics as a scientific study of how man uses the resources at his disposal to satisfy his wants and we have since been concerned to work out some of the implications of that definition. The classic definition of economics is that of Professor Lord Robbins in *The Nature and Significance of Economic Science* in which he says much the same thing but in more precise and elegant, if less readily comprehensible, language. 'Economics', writes Robbins, 'is the science which studies human behaviour as a relationship between ends and scarce means which have alternative uses.'

When the elements of this definition are taken together it becomes clear that economics is very much concerned with choice. Because wants are virtually unlimited, because resources are scarce, and because these resources are capable of alternative use, the whole of economic life is permeated with choice. If wants could all be satisfied by existing resources there would be no need for choice. If resources were not scarce there would be no need for choice. If each resource could only be used for a single purpose the question of choice would not arise. But where wants are unlimited, resources are scarce and can be used in different ways, choice is ever present.

Recognition of the circumstances giving rise to choice leads us to one of the most important concepts in economics, that of 'opportunity cost': the cost of A is the alternative B (or C and D) sacrificed in order to obtain A. For the individual or the family unit the cost of a dishwasher may be the freezer they have had to do without; the cost of some extra clothes in one month may be fewer cinema visits or some extra constraints in spending on food or drink. The concept of opportunity cost also applies at the national level: the cost of a new hospital may be the four new schools which could otherwise have been provided with the same resources; the cost of a bigger defence force is a lower level of social security provision; or, conversely, the cost of better social security provisions may be a weakening of defence.

It should be clear that what we are talking about is choice 'at the margin'. Resources will run to a number of items which do not require careful assessment. Some food, some clothes, will be purchased by the household. But at the margin of resources the difficult choices have to be made and opportunity cost comes into its own. Similarly with the nation. The choice is not all hospitals and no schools or vice versa, it is which of the two to choose as we approach the margin of resources.

But, the reader may say, I can accept opportunity cost for the individual or household with its strictly limited resources. I can even accept that it may be true for the nation if we have full employment. How can it be true for the nation in the early 1980s when there are large numbers of people out of work and much capital idle or under-utilised? With unemployed resources why, at the margin of state expenditure, does the choice of more hospitals imply less schools? Why can we not have both? As the teacher might say, 'That is a good question', to which we can only give a provisional answer at this stage. Had the question been put twenty years ago almost all economists would have replied that if there were high unemployment we could have both more hospitals and more schools, but that the unemployment was a temporary situation, an aberration, and that once a full employment equilibrium was reached, opportunity cost would come into its own again. But twenty years ago we did not have the combination of high unemployment and high inflation that has characterised the last ten or fifteen years. Even today most economists would say we could have more hospitals without sacrificing schools, but many would add (and this is the Conservative Government's view) that to pursue both would generate a higher rate of inflation. If reducing the rate of inflation is seen as the overriding objective, then compatible with that objective, at the margin of State expenditure, the choice between hospitals and schools remains.

Our analytical definition of economics provides the touchstone about what subjects come within its purview and indicates the approach to these subjects. Thus the economist is concerned with unemployment not because it is a soul-destroying evil (which it is), but because it represents the uneconomic use of resources. He is concerned with industrial organisation because different industrial structures represent different ways in which the factors of production are utilised to satisfy wants and some ways may be more economical than others. He is concerned with the size of firms because, for particular products, some sizes may yield economies, generating lower average costs. He studies international trade because each country can satisfy more of its wants by trade than if it was self-sufficient if each concentrates on producing those goods in which it has relative cost advantages.

Our examples do not need to be restricted to nations and firms. We have earlier given illustrations of economics at the individual and family level. Equally, we could look at voluntary organisations. The activities of a football club, an amateur dramatic society or a church can all be approached from the economist's viewpoint. Each organisation has some objectives or ends its members seek to promote. Each has scarce resources, of manpower and equipment, with which to achieve these objectives. The economics of the club or the church is the study of how these resources are used to achieve the ends and how, perhaps, the ends could be more fully attained by using the resources more economically.

In a study of the economic structure we shall be concentrating our attention on the 'macro' level, on taking a broad look at the framework of the economy. But this should not lead us to forget that there is an economic component of almost every kind of human activity we can mention.

2

THE NATIONAL ECONOMY

The national product

Probably no initials appear more frequently in press reports on the national economy than 'GNP'. Perhaps most newspaper readers know that the initials stand for Gross National Product. Certainly very few will appreciate its full implications.

Broadly speaking, GNP is a measure of the volume of goods and services available to the citizens of a country over a period of time, usually a year. At the beginning of the year in any country there will be in existence a large stock of equipment or capital of all kinds – buildings, machines, materials, semi-finished goods. There will also be a large area of land used for cultivation, building and other purposes. As the year gets under way so Labour, workers with all kinds of skills, will get to work with the other factors of production, Capital and Land and produce a stream of goods and services.

A large part of the goods and all the services are single-use consumers' goods like food, which once used are all used up. Others are durable-use consumers' goods, like washing machines, irons, refrigerators and clothes, which can go on being used for a considerable period of time after purchase. Other goods are producers' goods, which will be used as capital to help to make consumers' goods. Producers' goods can also be divided into single-use and durable-use. The single-use are materials and semi-finished goods and using them up means passing them on to a different stage of production. The durable-use producers' goods include machinery, factories and office blocks.

The purpose of the whole activity is consumption – a term the economist employs to cover the use and enjoyment of goods and services of all kinds, not just food. The size of this flow of goods and services determines the living standards of the community. There is, however, an important choice to be made about the composition of output. The larger the proportion going to consumers' goods the higher the standard of living in that year. On the other hand the larger the proportion going to producers' goods the higher the standard of living in future years. At the end of the year there will be a stock of goods of all kinds, but mainly durable-use goods, both consumers' and producers', and this stock is available to help to satisfy consumers' wants in the next year. The difference between the stock of capital left at the end of the year and that in existence at the beginning of the year represents the net addition to capital over the year, called net investment – which could be a negative figure if the community was partly 'living on its capital'. Gross investment is the total addition to capital before allowing for the capital used up during the year.

In attempting to measure the flow of goods and services to determine the total consumption and net investment we must be careful not to double count. For instance, we should not count as part of output both bread and the wheat and the flour which went to make it. The wheat and flour are embodied in the bread and in counting the bread we are taking in their value. In other words, we are only concerned with the value of the final products for consumption or investment.

Measuring national output

In describing the national economy we have so far concentrated on product or output. In order to obtain a measure of the value of output we can follow any of three routes: output, income or expenditure. Provided each is defined appropriately and our statistics are complete, we can derive the value of national output from any of these three: in other words, national income equals national output equals national expenditure.

This equality is fundamental to an understanding of an economy and we need to explore it further. In fact we have already touched on it when we said that the limit on an individual's income was the amount of the productive services he could supply, whether of Labour or Capital, and the value society placed on them. These services represent his contribution to the national product in exchange for which he obtains an income which, for convenience, is in the form of money, which he then converts into goods and services by spending. The goods and services he buys won't be the same goods and services that he helped to produce (though they might include some goods he helped to make) but, subject to some qualifications which we shall shortly make, they should equal them in value.

For this equivalence of income, output and expenditure to hold good for the national economy we have to recognise that some income does not accrue directly to individuals. In the United Kingdom such income is of two kinds. Charitable trusts receive income from property, which is then spent by the trustees. Also, income occurs in the form of undistributed profits – receipts of companies from selling products which, strictly speaking, belongs to the shareholders but has not been distributed to them and which they cannot get their hands on and call their own. If we should fail to take these forms of income into the reckoning then the income column would add up to less than the value of output.

In totalling the value of output we can follow either of two procedures. One is to add up the value of consumption goods and services plus investment in the way we have already indicated. Another is to follow through all stages of production – primary industry, manufacturing and distribution – and total the 'value added' at each stage of the productive process. Value added is the difference between the value of the raw materials and services a firm buys in and the value of the product it sells. It represents the contribution of each firm to the value of the product it sells.

On the expenditure side there are also two procedures which may be followed. One is to look at the purchase of all the goods and services which comprise output including new equipment provided by firms. Another is to take the value of consumption goods and services plus saving. It may seem something of an oddity to count 'saving' as expenditure; but saving should be thought of as the opposite of consumption rather than the opposite of spending. Many forms of saving do involve 'spending', e.g. the purchase of a government saving certificate or bond, paying the premium on an insurance policy, or buying a house. Only if saving takes the form of bank or building society deposits or stacking money away in a stocking under the bed does it not involve 'spending'.

16 *The economic structure*

In practice, the way we derive the somewhat imperfect information to enable us to estimate the size of the national income, output and expenditure is from statistics collected by government. The main source of income statistics is data derived from income tax returns. Censuses of farming, production and distribution enable us to fill out the output column. Expenditure data comes from a variety of sources including family expenditure and household surveys.

Complications

Having presented this account of national income and output in relatively simple terms, we now need to recognise that, regrettably, there are a number of complications and limitations to the picture.

1. Money measurement Implicitly we have already recognised that, in order to compute the totals of income, output and expenditure, we have to convert the goods and services into money values. But the reality behind the money values must not be forgotten. 'Real' income is the goods and services which a person (or a nation) can enjoy over a period of time without living on capital. Money income is simply the money amount per period of time in exchange for services rendered. Real income is the more important. If money income doubles but real income doesn't change, we can only enjoy the same volume of goods and services as before. We are no better off. On the other hand, if real income increases we are better off even if money income were to remain the same. The first situation implies that prices have doubled; the second, that they have fallen. Money is indispensable both for affecting exchanges and as a unit of account; but, as we shall examine in more depth later, its use is not devoid of problems.

2. The missing economies A fundamental limitation of our picture is that national output as measured in the statistics of governments does not cover the whole of the goods and services produced and enjoyed in a nation. It covers only those which satisfy two conditions: (1) they are exchanged, i.e. bought and sold; and (2) that exchange comes within the purview of the State.

In other words, official figures of national output only relate to what may be called the formal economy. There is, in addition, the

The national economy 17

household economy and the informal economy. These three economies may be thought of as overlapping circles of economic activity as portrayed in Fig. 1.

The household economy consists of the whole range of goods and services supplied by the household for its own use. Thus a man may grow vegetables for his family to eat and flowers for the house; his son may mow the lawn and do the decorating; his wife may cook and clean. All of these activities generate goods or services which might have been paid for but weren't: he could have bought vegetables and flowers from the shops; employed a gardener and a painter; bought the services of a cleaner and a cook. Had he done

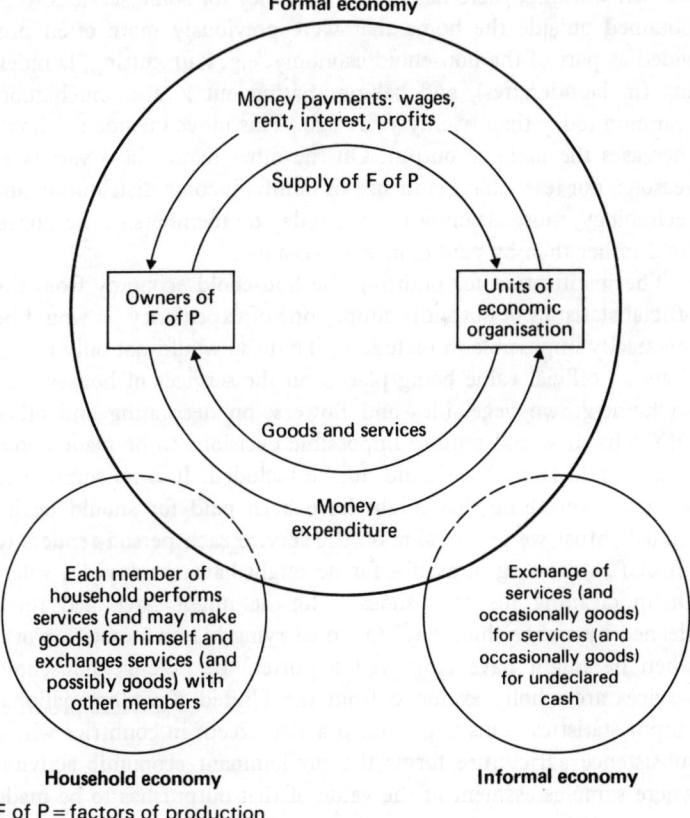

F of P = factors of production

Fig. 1 The three economies

so all these items would have appeared in the statistics as part of the national output; as it is, they do not so figure.

The frontier between the household economy and the formal economy which alone figures in the national statistics is a moving one and changes in the frontier may create misleading impressions. Thus, if fewer women bake their own bread and instead buy it in the shops, the national output goes up; while if a man marries his housekeeper the national output goes down, even if she now cooks and cleans with more will and to more effect than before because her heart is in the job. Contrary tendencies may be distinguished over the years. On the one hand, especially with more married women working, there has been a tendency for some services to be obtained outside the home that were previously more often provided as part of the household economy, e.g. hair cutting, laundering (in launderettes), and baking. Eating out is also much more common today than twenty years ago. This move outside the home increases the national output. On the other hand, for a variety of reasons, not least changes in size of family, income distribution and technology, more cleaning is done today by members of the household rather than by paid domestic servants.

The justification for omitting the household economy from the official statistics of output is simply one of expediency. It would be practically impossible to include it. To do so would not only necessitate an official value being placed on the services of housewives, on home-grown vegetables and flowers, on decorating and other DIY jobs, it would require impossible decisions to be made about what should and what should not be included. It is no solution to say that everything that might have been paid for should be included. Must we put a value on the service each person renders to himself in dressing himself – for he might have employed a valet? Or in cleaning his own shoes – for he might have had them cleaned by a shoe-shine boy? Or in carrying his own case to work, when he might have employed a porter? In practice household services are wholly excluded from the United Kingdom national output statistics. This is general practice except in countries where subsistence agriculture forms the predominant economic activity, where some assessment of the value of that output has to be made if national output figures are to be meaningful.

It matters less in the United Kingdom where the size of the

household economy is smaller. However, the omission becomes important if there is some major shift between the formal and household economies. For example, in wartime, many more people may 'dig for victory' by growing their own food. More recently, we have become aware of a possible shift towards the household economy as a result of taxation. Reducing expenditure by DIY activities is an attractive alternative to increasing income if income is heavily taxed. For example, suppose a university lecturer could earn £600 from writing articles in the time it would take him to paint his house and that he is indifferent between these activities as leisure occupations. Suppose also that it would cost him £500 to pay a professional decorator. In the absence of income tax the preferred alternative would clearly be to write the articles and employ the decorator. But if one-third of his income goes in tax, then he is £100 better off by painting the house himself.

The informal economy consists of barter exchanges outside the household or of exchanges for money which do not appear in official income statistics. Thus a group of men with different skills may join a syndicate to build houses for themselves and each other in their leisure time. The plumber amongst them will do the plumbing in all the houses and in exchange gets the plastering, carpentry and so on done in his house. No money changes hands between them. Similarly, parents of young children may form a local babysitting club in which each member undertakes to sit four times a month for others in exchange for four occasions when others babysit for them. Such arrangements might be regarded as the extension of the principles of the household economy to a wider circle of neighbours and friends. No laws are broken with these barter deals, but we are approaching a difficult borderline. Thus someone might give a lecture without asking or expecting a fee, and be given a bottle of wine in gratitude. There was no contract to pay and the wine can properly be treated as a present which the lecturer is under no obligation to declare to the income tax authorities. But if, by choice, in place of a fee he received a case of claret, then not to declare that is a tax evasion with the result that an economic activity which should have appeared as part of national income and output does not so do. The legitimate shades into the illegitimate and the informal economy embraces the 'black' economy. The black economy comprises activities which ought to be under the cognizance of the authorities but are not, because the object is tax evasion. The scope for such evasion arises mainly from self-employment: the trader who, in the course of his work, under-declares his income either by under-stating his receipts ('putting his hand in the till') or over-stating his deductible expenses; and the 'moonlighter' who does jobs in his spare time for cash which he does not declare as income.

Concern has recently been expressed in the United Kingdom about the black economy and many people believe that, for a number of reasons, including high taxation, unemployment and a decline in tax morality, it is growing. A recent chairman of the Board of Inland Revenue, Sir William Pile, expressed the view that it was 'not implausible' that the black economy might be equal in size to about 7.5 per cent of the national income. The informal economy would, of course, be larger than this for not all the informal economy is illegitimate. But, legitimate or not, the activities of the informal economy are either not reflected at all, or are not fully reflected in the official figures of national output.

In short, to sum up this section, GNP or the official figures of national income and output do not cover the whole of economic activity; and changes in GNP could reflect not changes in economic activity but a shift from the formal economy to the household and informal economies or vice versa.

3. Government activity The relationship between income, output and expenditure is considerably complicated as a result of government activity, principally through taxation and transfer payments. Government taxes are levied in order to pay for communal services like the defence forces, education in state schools and the National Health Service. These taxes can be thought of as similar to a subscription where the taxpayers, as it were, club together to provide these services for them all, just as many of them might co-operate and pool some of their resources to found a tennis club. There are, of course, important differences: tax is a compulsory levy, not a voluntary subscription, and the tax will not be at a uniform rate for all taxpayers as it usually is for all club members. But the analogy is close enough and, interpreted in this way, where economic services are provided through the medium of the State, no problem arises of any divergence of income, output and expenditure. However, where the tax payments are to effect transfer payments there is a problem and, indeed, we touched on it earlier (p. 6). Transfer payments are income-type payments and receipts, like social security benefits, which do not generate a corresponding economic activity. Taxation to pay for transfer payments creates no communal services for all but simply transfers income from the taxpayer to the social security beneficiary, who can proceed to spend it on personal goods and services. In aggregating incomes to obtain the national income, if we included both the taxpayer's pre-tax income and the transfer incomes, we would find the income total exceeding both the value of total output and total expenditure. In other words we would be double-counting. To retain the equivalence of total income, output and expenditure we must either exclude from the taxpayer's income that component of tax which pays for transfer payments or, more conveniently, simply deduct transfer incomes from the income column.

A further problem arises in relation to taxes and subsidies on goods and services. Taxes minus subsidies on goods and services can be conveniently referred to as 'net' indirect taxes. The presence of net indirect taxes means that market prices differ from costs of production or factor costs. If net taxes are positive (taxes are more than subsidies) then the sum of the expenditure column will exceed that of the income and output columns. To retain the equality we need to deduct net indirect taxes from the expenditure column so that all three are measured at factor cost.

4. International trade Complications about the equivalence of income, output and expenditure also arise where a nation engages in foreign trade. Let us start with the simplest situation of no foreign lending and borrowing, past or present. This implies that the value of exports and the value of imports are precisely equal. In this situation income, output and expenditure remain equal in total value, but the composition of the expenditure column changes: imported goods and services, which have been exchanged for exported output, now figure in expenditure.

Suppose now, for the first time, a country's exports exceed the value of its imports. It is said to have a surplus on its balance of payments, which can only happen if the importing countries are given the means to pay for the excess of imports. In other words, foreign lending has taken place. In this situation the equivalence of total income, output and expenditure is maintained provided that foreign lending is included in the expenditure column as part of saving.

We now come to the most complicated case – where foreign lending and borrowing has been taking place in the past. Countries which have been net lenders are called creditor nations and countries which have been net borrowers are called debtor nations. Let us illustrate by taking a creditor nation like the United Kingdom. Included in the income of UK nationals will be some net income from overseas assets. This income can be spent, so the expenditure column will include some imports which have not had to be paid for by exports or some further overseas lending. The income and expenditure columns will balance but will be larger than the output produced in the nation. In effect, part of the output corresponding to the income column is being produced overseas. We can therefore maintain the equality either by adding on the net value of the overseas output produced by UK nationals, or, what amounts to the same thing and is more convenient, by adding to output the net income from overseas assets.

The concept of domestic output or product gives us another measure of output, different from national output, which is a better measure of economic activity if we want to look specifically at what is produced at home. Domestic output will, of course, be smaller than national income or output for a creditor nation and larger than national income or output for a debtor nation.

5. National output as a measure of welfare National output, especially when divided by the total population to give a measure of output *per capita*, is often used as an indicator of economic welfare; and changes in national output over a series of years are used to indicate economic growth (or decline). GNP (or one of the other measures of national output) is probably the best single measure of economic growth and economic welfare, but it has many imperfections, some of which we have already considered. We need to measure national output in money terms but, in any comparison over a period of years, we must allow for changes in the value of money if the comparison is to be meaningful. The usual method is to value all the goods and services of every year in terms of the prices ruling in one particular year. Again, we have noted that official figures do not cover either the household or the informal economies and any switch between them and the formal economy could distort comparisons. There are also other problems.

Some of the components of national output as measured in official figures represent costs rather than benefits. Consider some of the costs of transport. Many workers, especially in the big urban areas, incur very considerable costs – motoring, bus, train or underground – in travelling regularly between home and work. In logic, what they pay for this transport might best be regarded as a cost of producing whatever goods and services these workers help to provide. However, in the national accounts the cost of such travel is treated as a consumption expenditure, which is assumed to constitute part of living standards.

Another not dissimilar situation arises with pollution. A firm may pollute a local river in the course of production. If the government incurs expenditure in removing the pollution, this additional expenditure is regarded as providing consumption benefits for the community even though it simply neutralises the effect of the pollution caused by the firm.

Another influence on real growth and welfare, hidden in the statistics, is change in the quality of goods and services. Apparently similar goods may vary in quality over time, e.g. a newspaper may be halved in size but it still appears in the figures as a newspaper. One such change has been brought about by supermarkets. In the pre-supermarket era, when we bought groceries and the like we could expect shop assistants to take them from the shelves, pack them for us and deliver them to our home. Now we collect them from the shelves ourselves, queue at the checkout, pack them ourselves and carry them home ourselves. This change in the quality of service going with goods is not registered in the national output statistics.

There are particular pitfalls if one is attempting to measure changes in economic welfare by comparing changes in output per head. Besides the limitations already mentioned, there is a problem about the composition of government services. For example, there might be a big increase in resources devoted to defence. There is an important sense in which the community may benefit from this expenditure, but it is not unlikely in these circumstances that national output might have gone up while personal welfare – the value of goods and services entering personal living standards – had gone down. Another important item in any comparison of welfare is the amount of leisure people enjoy. The national output statistics tell us nothing about this. Thus if output *per capita* has fallen but so have hours of work, so that workers have more leisure, it is not clear that economic welfare is less.

These same difficulties occur in using national income or output statistics in international comparisons of welfare and are accentuated by others. For example, how do you allow for the fact that the inhabitants of a country with a warm climate all year round do not need to use resources for domestic heating?

Concepts and definitions

This chapter has contained many important concepts and definitions, sometimes implicitly rather than explicitly stated. In this final section we attempt a brief recapitulation and also attempt to make the definitions explicit.

Income, output and expenditure, when appropriately defined, are related to each other in a very direct and fundamental way. That relationship is complicated but not destroyed when we take account of government action and international trade.

In terms of official national income and output statistics, in effect we define output as the flow of goods and services produced over a period of time and entering officially into exchange. This definition omits the output of the household and the informal economy. Consumption is the satisfaction of wants through exchanges which are officially recognised. Investment is the addition to capital over a period of time and net investment is investment minus depreciation or capital consumption.

There are a number of different concepts of national output or income of which GNP is only one. These different concepts arise from (1) differences between gross and net: the deduction of depreciation from gross output gives net output; (2) differences between 'national' and 'domestic': domestic product takes no account of net income from foreign assets (which may be negative); and (3) differences between factor cost and market prices: the market price

TABLE 1 *Concepts of national output*

Net DP at FC + Net income from foreign assets = Net NP at FC
Net NP at FC + Net investment = GNP at FC
GNP at FC + Net indirect taxes = GNP at MP

Where D = domestic, N = national, P = product or output, FC = factor cost, and MP = market prices.

Each concept may be more useful than the others for particular purposes.

---

measure includes the effect on prices of taxes and subsidies on goods and services. If we take a creditor nation in which taxes on goods and services are more than subsidies, we can show these relationships in summary form starting from the concept of output which gives the smallest figure, and proceeding to the largest (Table 1).

Finally, whilst national output and income statistics may be the best indicator of economic growth and economic welfare, they suffer from a number of deficiencies.

In this chapter we have presented a picture of the national economy and of national income, output and expenditure; but we have said nothing about the crucial questions in any economy: who determines what shall be produced and how it shall be produced and who shall get the product. The basic economic problem, of trying to satisfy virtually unlimited ends with scarce resources that can be used in different ways, is the same for any form of socio-economic organisation, but the method of resolving it may differ. In the next two chapters we look at two alternative solutions representing opposite extremes. In the first case State activity is minimal; in the second it is total. Neither of these situations exists in practice in that extreme form, but an account of their principles of operation and their limitations provide standards of reference which help us to understand the mixed economy of the United Kingdom.

3

SOLVING THE ECONOMIC PROBLEM: THE MARKET SYSTEM

Markets and the price system

In an economy with minimal State activity (the provision of a currency system and the maintenance of law, order and defence services) the crucial economic issues of what shall be produced, how it shall be produced and who shall get the product, are resolved through the operation of markets and the price system.

When we speak of a 'market' we tend to think of a particular place, like the splendid market at Leicester where a large number of stalls offering excellent value for money are assembled in a market hall. But a market to an economist means something more than that. It is not so much a place as a whole network of dealings, which might be by telephone. The criterion is that buyers and sellers must be in such close contact with each other that a change of price in one part of the market affects the prices paid in every other part. Some markets may be very localised; others are extended worldwide by the aid of telephone and Telex links. Further, markets are not confined to dealings in goods and services but extend to the factors of production. Thus there is a labour market, where labour services are bought and sold; a land market; markets for materials; and capital markets (which are concerned with the borrowing and lending of funds by means of which capital goods can be acquired).

In markets, goods and services and factors of production are bought and sold at a price. The price of labour is called a wage-rate or salary; the price of capital is the payment lenders require from borrowers – the rate of interest.

The price system is the term used to describe the whole inter-related structure of prices and markets. It is through prices and changes in prices that the economic problem of scarce means and unlimited wants is resolved. Because incomes, output and expenditure are so inter-related it is difficult to know where to break into the circle to explain the workings of the price system. But let us start with the 'what', 'how' and 'who' questions with which we began the chapter.

What shall be produced is decided through the interaction in markets of the decisions of entrepreneurs, who organise the factors of production, and of consumers, who derive income from selling their labour services or lending their property. Thousands of entrepreneurs buy the services of the factors of production and put products on the market at a price to cover the cost of the factors and give them a margin of profit. The millions of consumers, each making his own decision in the light of his income and the prices ruling, decide whether, and how much, to buy. If the consumers are not sufficiently attracted by a particular product to clear the market at the offered price, the price will fall. Then production is cut back and, in the longer run, the highest cost firms may go out of business. The reduction in supply will check the fall in price and may reverse it somewhat; a price equilibrium is reached in that product market which gives existing firms a sufficient profit margin to make it worthwhile staying in business, but not so high a margin that new firms are attracted into that line of production. The resources employed by the firms which have gone out of production will seek employment elsewhere. If consumers want more of a particular product at the price it is being supplied, its price will rise. It will be profitable for existing producers to make more and for new producers to move into that market.

The product markets and the factor markets are closely linked. Thus if the price of a product rises, giving a larger margin of profit to entrepreneurs producing it, they will wish to expand production and new entrepreneurs will seek to enter the market and both will be prepared to pay more for the factors of production needed for that particular product. So the wages of labour in that industry rise and more workers are attracted to it; and the prices of machines used in that industry also rise and it becomes more profitable to make those machines; more workers are needed for that purpose, too, and the wages and salaries of such workers increase. And so on. Conversely, the wages and salaries of workers making goods for which demand has fallen will decline.

In the market economy new products are pioneered by entrepreneurs in the expectation of profit but it is the consumers who, acting individually, in aggregate determine whether that product will continue to be produced or not; they exercise a continuous right of veto.

Because the price system applies also to the factors of production, prices govern not only what shall be produced but how it shall be produced. Product prices, as we have indicated, react back on factor markets, and prices in factor markets are also influenced by supply considerations such as the exhaustion of a particular raw material or the temporary lack of a particular labour skill or machine. Prices in the factor market thus determine the 'factor mix'.

Because factors of production can be used in a variety of ways, a given level of output can be produced by many different combinations of factors of production. One natural raw material may be substituted for another or a man made material for a natural material; labour of one skill can be replaced by that of another; a new type of machine can replace another which is more costly or less efficient; machines may be substituted for men or vice versa; labour can even to a small extent be substituted for raw materials, e.g. if a particular fibre used, say, in making cloth is becoming scarcer, then more labour or labour of greater skill may be used in the various stages of production to cut down the proportion of the material which finishes up as waste. The scope for factor substitution will vary according to the nature of the product but in some measure factor substitution will be possible for all products. The entrepreneur, in pursuit of profit, will select that combination of factors of production which will produce the planned output level at least cost. In this way the 'how' question is answered.

The outcome of the operation of markets also determines who gets the product. The price of labour is the income of the wage-earner; the price of capital is the rate of interest; the price for the use of land is rent; and the entrepreneur receives profit for risk-taking. The size of these rewards reflects the relative scarcities of the many different factors of production at any time. A person's income is the value which society puts on the supply of the factors of production he offers on the market.

Merits of the price system

From this description of the operation of the price system we can draw out a list of its main merits. It has a series of inter-connected functions.

An information system

Prices provide a vital information service. They act as signals to consumers, to entrepreneurs and to the owners of the factors of production. Consumers compare the prices of different goods and services in deciding what to buy. Changes in product prices are an indication to entrepreneurs about what to produce in the future. The prices of the factors of production guide entrepreneurs in determining what methods of production to employ, while at the same time indicating to the suppliers of labour, land and capital which economic employments are more financially rewarding and which should be avoided.

A rationing system

The price system rations out the available supply of goods and services among consumers. The term 'ration' tends to conjure up the picture of a siege economy, but this is misleading. *Some* system of rationing is essential whatever the form of socio-economic organisation, because the available goods and services always fall short of the wants we should like to satisfy. The virtue of the price system is that it distributes the scarce goods and services in such a way that each individual, within his budget limitations, maximises his satisfaction. Each can buy that combination of goods and services which he expects will give him the most enjoyment or (to use a term much employed in economics) 'utility'. Contrast this method of distribution with a system of physical rationing. Suppose I like eggs but not bacon. With rationing by physical quantities of the goods I get a quota of both. By sheer good luck I may be able to find someone whose tastes are precisely the opposite to mine, with whom I can exchange bacon for eggs. This is, however, unlikely and, in any case, the structure of individual tastes is much more complex than simply absolute likes and dislikes. Even if I liked all the rationed products I almost certainly would prefer them in different proportions to what is on offer. Prices as a rationing mechanism offer the maximum flexibility.

30 *The economic structure*

An efficiency mechanism

Under a market system prices reflect the real resource cost of a product and ensure that scarce resources are economically used. For example, suppose that a particular raw material is becoming scarce. Its price will rise and as a result entrepreneurs seeking profits will find it worthwhile to substitute other factors of production for the factor which has risen in price whenever the substitution can be easily and cheaply effected. Thus the particularly scarce material will be left for those uses where substitution is impossible or expensive.

A coordinating mechanism

Prices serve as a remarkably subtle way of coordinating the consumption and production decisions of millions of different individuals. In economics it is a truism that 'everything depends on everything else'. As we have seen, changes in consumer behaviour in one product market affect other product markets and a variety of factor markets. The price system is an automatic and pervasive coordinating mechanism to enable the community's wants to be met efficiently.

Harmonisation of interest

It was the way the price system coordinated individual and communal ends which led Adam Smith to eulogise about the 'invisible hand'. Each individual, he maintained, in following his own self-interest, was automatically serving the interest of the community, as though led by an invisible hand to pursue an end which was no part of his intention. Suppose people wish to eat more fish. Then the price of fish will rise. In response to that rise trawler owners seeking profits will work their boats for more hours, pay overtime rates to their fishermen and seek to employ more fishermen by offering higher wages. Trawler owners, established fishermen, new fishermen, all act from self-interest; the effect is to meet the community's demand for more fish.

A mechanism of delegation

One of the attractions of the price system is that it decentralises decision-making. Decisions are made by millions of individuals, each acting in the light of the limited amount of information neces-

Solving the economic problem: the market system

sary for him to make his own decision. Decision-making is disseminated through the community and economic power is dispersed.

Limitations and deficiencies of the price system

In fact the picture of the price system we have presented is idealised. While there is a vital element of truth in all the merits claimed for it, the price system does not work in the ideal way described; even if it did, it would still suffer from one inherent defect. In practice, five main limitations or positive deficiencies can be identified.

Goods with diffuse and indivisible benefits

There are some goods or services which give a benefit which is indivisible and diffuse. The provision of a police force, a legal system, a currency system and defence forces all come into this category. The individual cannot buy his own little bit of these services in markets. They have to be provided communally. It would be possible, in theory, for a small community to club together and all agree to contribute to provide these services, thus maintaining the voluntary principle. However, unless unanimity on expenditures and contributions could be agreed, it would be impossible to stop 'free-riding', for these are all services to which the exclusion principle (that those not contributing can be excluded from enjoyment) does not apply. In practice these services have to be provided by some sort of compulsory levy.

This limitation of the price system was recognised in the first paragraph of this chapter when, in describing the market system, we acknowledged the necessity of minimal State activity (p. 26).

However, the problem of goods and services offering a diffuse benefit is not confined to those which *must* be provided by the State. There are others where it would be possible to charge users, so that the good or service *could* be sold in markets, but where it would be highly inconvenient or expensive so to do. Consider the provision of roads in towns. While it is perfectly feasible to levy a toll for the use of trunk roads, it would be highly inconvenient to impose a charge on users of the many intersecting roads in towns. Similarly with the paving and lighting of towns. It is much more convenient and less costly if these services are provided and maintained by a communal agency and the cost met by a compulsory charge on local inhabitants.

Externality effects

The situation of goods with diffuse and indivisible benefits, where the exclusion principle cannot easily be applied, merges into the situation where, although there are individual benefits which be clearly identified and priced, there are also 'spillover' or 'externality' effects. To put the same point in another way: in both production and consumption there may be social costs or social benefits over and above private costs and benefits. A private cost or benefit is also a social cost or benefit, for the individual is a member of society. But other people, who are not parties to the activity in question, may suffer costs or enjoy benefits. A series of examples will make the point.

Consider first the production side. The way a particular good is produced may cause river pollution from the disposal of waste products, air pollution from black smoke, or noise pollution for lack of sound-proofing. These are costs to society over and above the costs of production taken into account by the entrepreneur. On the other hand some production activities may diffuse social benefits; thus the good farmer, in keeping fields and hedgerows trim and paths clear, is conferring social benefit by increasing the beauty and accessibility of the countryside.

On the consumption side, heavy drinkers may generate costs to society from riotous behaviour, requiring more expenditure on law and order; or, more seriously, may cause road deaths by drunken driving. But some consumption activities generate social benefits. Thus people who seek prompt medical treatment for infectious and contagious diseases benefit those who might otherwise have caught them; and the increase in human knowledge arising indirectly from the consumption of higher education may bring widespread benefits ranging from improved medicine, reducing illness, to a better understanding of the economy, reducing unemployment.

The point about these externalities is that, in the absence of any State or communal intervention, market prices only take into account private costs and benefits. Where there are additional social costs and benefits the optimum output is not produced. Thus the price of the product of the polluting firm is set lower than if the costs of pollution were taken into account, and its output is larger than is justified. Similarly, in the absence of intervention the amount of private expenditure on health or education would be less than optimum.

In considering whether a divergence between social and private costs or benefits justifies State intervention and, if so, of what kind, several points should be borne in mind. First, almost every activity generates some social costs or benefits over and above private; so there must be a margin of tolerance. The man who neglects his front garden allows an eyesore to develop, which is a social cost; while he who tends his garden creates a social benefit over and above private benefit; but it would be absurd to suggest that the State should penalise bad gardeners and reward good gardeners. Secondly, unless the extent of a divergence can be measured with some broad degree of accuracy, intervention may do more harm than good by over-correction. Finally, the borderline between private and social cost is not fixed. By appropriate regulation (e.g. defining and banning unacceptable levels of pollution) what was previously a social but not a private cost may be converted to a private cost.

Monopoly and imperfect competition

The perfect functioning of the price system implies perfect competition – a situation in which in every market sellers are numerous and no seller, acting alone, is sufficiently large to have any significant effect on price. Price is fixed by the market; all sellers are 'price takers' rather than 'price setters'. Where elements of imperfect competition or monopoly creep in, the price system does not operate in the way we have described it. For example, suppose a monopoly develops in a particular line of production. It may be based on technical economies, which make it uneconomic for a new firm to enter the market except on a very large scale; or on control of the supply of a raw material essential to that product. If demand for the product increases its price will rise, but the price rise will not attract more producers into that market; and the monopolist himself may not increase production, or not by much, but rather make higher profits per unit. Then the price ceases to reflect real resource costs and the harmony of individual and communal interest is not attained. Similarly with a monopoly in the supply of labour. If a trade union controls all the supply of a particular kind of skilled labour essential to the production of a particular good or service, including control over new entrants, then an increase in demand for that type of labour, resulting from an increase in demand for the products it makes, will simply raise the wages of those workers without increasing their supply. Some increase in the supply of the product may result from factor substitution, but that increase will be less than would have occurred in the absence of monopoly.

Some types of advertising are aimed at conferring an element of monopoly power by inducing the purchasers to believe in the superior qualities of the product and creating a brand loyalty. Then the producers can have some control over their product price; within limits they will be able to raise it without frightening away their customers. On the other hand other forms of advertising, like the programmes at theatres and cinemas or the advertisement of a new product, increase the competitiveness of the economy by increasing knowledge.

Fluctuations in activity and employment

Experience has shown that the *laissez-faire* economy, the economy with minimum State activity, has only a limited stability: that it has generated fluctuations in activity and employment known as the 'trade cycle'. The classical economists argued that unemployment could not be other than a temporary phenomenon because 'supply creates its own demand'; the incomes received by the factors of production create the demand to maintain production at a full employment equilibrium. If all income was spent on consumption goods and services there would be no problem. The trouble arises from the relationship between saving and investment. If people save (i.e. refrain from consuming) part of their income but spend the savings on investment goods themselves, e.g. by buying a house, aggregate demand and therefore employment is maintained. But if they save and do not themselves invest, employment is only maintained if someone else invests (i.e. adds to the stock of capital) to the extent of their saving. In classical theory the rate of interest was regarded as the mechanism which kept saving and investment in line. If few people wished to borrow to invest, then interest rates would be low; saving would then be discouraged and consumption would increase. If the demand for investible funds was high, high interest

Solving the economic problem: the market system

rates would encourage saving and lending. The rate of interest was the price which equated the supply of saving and the demand for investible funds. Keynes showed, however, that the rate of interest could not be relied on to fulfil this function. If interest rates were low savers might not turn to consumption but simply hold off the market in anticipation of a future rise. In that situation aggregate demand would fall, activity contract and unemployment result.

Moreover, a situation in which a slump was anticipated reveals another contradiction to Adam Smith's optimistic assumption of the harmonisation of individual and communal interest. If workers expect unemployment they tend to consume less and save against a rainy day. If entrepreneurs expect a slump they cut back investment in new machines. Both actions reduce demand and generate unemployment. In these circumstances, in pursuing his own self-interest the individual is *not* at the same time pursuing the interests of society, but precisely the reverse.

Inequality in incomes

Even if the price system were operating perfectly, in accordance with the ideal model, it would still be open to objection on grounds of equity. The income people receive represents the price society puts on the services they supply, whether labour services or the loan of their property. Some people with the good fortune to be born with scarce natural abilities, like pop stars or international footballers, receive very high incomes, while others, lacking natural abilities, may receive low incomes. Income derived from inherited property accentuates the inequalities which arise from income from work.

Moreover, the efficient operation of the price system *depends* on the generation of income inequalities. Changes in incomes are the means of redeploying labour and capital. Thus, workers who may have spent years acquiring a particular skill suffer a heavy loss of income if new technology makes them redundant; they may have to take unskilled work as an alternative. Conversely, workers who have the good fortune to possess a skill for which the demand is increasing may obtain very high wages. It is in society's interest that changes in the demands for different products should generate a speedy redeployment of labour, but the changes in the price of labour to bring this about are inevitably inequitable. Such changes in the income of workers are quite unrelated to the merits or faults of the beneficiaries or sufferers.

The future and the philosophy

These are not the only objections which can be made to the market system. There are also more subtle and debatable objections, to some extent interconnected and based on value judgements, concerning both its future and the philosophy behind it. The market system implies a consumer-oriented society. The objective or rationale of the market system is the satisfaction of the wants of consumers. Some would prefer a more socially-oriented society. Because markets reflect the demands of a *present* generation it can be argued that insufficient notice is taken of the needs of future generations in, for example, the rate of utilisation of non-renewable natural resources. Again, the philosophy behind a market economy is that of reward in relation to contribution; others may prefer a philosophy which attempts to relate reward to need. Moreover, some objectors to the market system dislike its motivating force – self-interest. Entrepreneurs in pursuit of personal profit, workers in pursuit of higher wages, consumers seeking to maximise utility – these are the mainsprings of the system. The profits and the higher wages may be partly used for philanthropic purposes and the utility function of the consumer may embrace the well-being of his neighbour. Nonetheless, some people perceive the market system as the embodiment of selfishness.

Let us look now at the opposite extreme, central economic planning, to see what that has to offer.

4

SOLVING THE ECONOMIC PROBLEM: THE CENTRALLY PLANNED ECONOMY

How central planning works

Central planning has a strong and immediate intellectual attraction. How much better that man should control his own destiny, it may be said, than be the plaything of 'blind economic forces'. Who could be against planning, defined by the dictionary as the 'exercise of rational foresight'. Who indeed? But the real questions are on what scale the planning takes place and who carries it out. In the market economy entrepreneurs plan, the owners of the factors of production plan, consumers plan; but there is no body which seeks to establish a national economic plan. It is the implications of this latter kind of plan, as a means of resolving the economic problem of scarcity, that we must now examine. A centrally planned economy is often referred to as a 'command economy'.

In a command economy an economic dictator, or a committee, working through a planning agency, determines what shall be produced, who shall produce it and who shall get the product.

Consider what an extreme form of command economy, without markets, would mean. In theory a huge balance sheet would need to be drawn up. On the one side would be all the resources available – the different qualities of land and their current uses, all the different kinds of labour and their locations, and the tens of thousands of forms of capital goods and materials. To set against this list of factors would be the goods and services the planners would like to see produced on the basis of some not wholly unrealistic assessment of what was possible. From past experience ratios (based on the productivity of the different factors of production) would be available which would indicate what volume of goods and services might be expected from what factor inputs. The list of desirable ends would exceed the production possibilities and the planners on behalf of the whole community would have to make the choices and decide what was to be produced.

The planners would also have to determine what methods of production to use. To some extent this decision could be delegated downwards to individual enterprises, but the enterprise must be given its production target on the one hand and its allocation of the various factors of production on the other, so its scope for factor substitution is limited. Suppose the planners wish to effect some change in priorities and redeploy labour; in the absence of markets offering a personal incentive to workers to move, how is the change brought about? The answer is by persuasion and direction. Volunteers may be sought on the grounds that the change is 'in the national interest' or 'for the good of the community'; but underlying the persuasion, to be used as necessary, would be the power of compulsion.

The planners, too, determine who shall get the product. They decide the distribution of income; if markets were completely absent then the allocation would be by means of physical rationing. There would be no money income but only an allocation of real income.

One has only to pause a moment to think how completely impossible it would be to run a large and sophisticated economy on this basis. In one respect the problems may be less difficult than part of our outline implied, in that planning takes place from an existing situation; production methods and outputs do not have to be determined from scratch and the main decisions are about year-to-year adjustments. But in every other respect our brief description vastly over-simplifies. Some idea of the enormous complexity of the central planners' task is given by the following quotation from a recent publication.

The statistical impossibility of detailed control of the economy from the centre was implied in the warnings about the future of the Soviet system uttered in 1964 by Victor M. Glushkov, then head of the Soviet Union's research programme in cybernetics. . . . He estimated that, even if higher speed computers were used, performing 30,000 operations per second, it would require a million computers working without interruption for several years to plan the entire economy. Nor is this a surprising calculation given that, according to Professor S. Warren Nutter, the total number of economic relationships within the Soviet Union approach several quintillion. Furthermore, since the economy is forever changing, the data fed into the computers

would require continual revision, with the result that the planners could never catch up with events. Consequently, even Glushkov's alarming calculation underestimates the magnitude of the task facing Soviet planners.
(Elst 1981, p. 17)

In practice, central planning necessarily becomes restricted to a comparatively limited and selective range of products and there is a delegation to regional economic councils and enterprises; this, however, creates its own problems of communication and coordination.

To do completely without labour markets and without markets for consumer goods and services is also quite impossible for an advanced economy. It would mean a system of barter, in which people were paid by the state in the form of a ration of food and clothes and so on. It would be possible for people to be given a basic weekly ration of those items which are frequently and regularly consumed, although at the cost of a considerable loss of consumer satisfaction, as it would take no account of individual tastes. But clothing, consumer durables and services (like hair-cutting) could not be given out at the end of each week. Some arrangements would have to be made, like the payment of workers in vouchers, which could be saved up for the larger items, and which would enable the workers to get their hair cut at a time which suited them. But this procedure implicitly introduces markets; and, especially if the vouchers could be exchanged against a range of goods or services, they would effectively be money.

In fact no advanced economy exists which doesn't use markets to some extent. In the Soviet Union and Eastern Europe, Cuba and China, consumer goods are sold in markets and labour is hired in markets; but prices do not play the signalling and allocating role they do in market economies.

Merits of command economies

Fairer income distribution

Perhaps the main advantage which can be claimed for a command economy is that income distribution can be determined by criteria other than those of the market. If changes in the price of labour are not the means by which the deployment of labour is determined, then incomes can be more closely allied to criteria such as 'from each according to his ability, to each according to his need'. Income inequalities can be reduced. There will be no need to pay outrageously high salaries to people who, by good fortune, possess scarce talents in popular demand, like pop stars or outstanding footballers. The planners, not the market, determine income differentials.

No unemployment

In a command economy where all workers are on the State's payroll, no-one is 'unemployed', so there are no reductions in income arising from loss of job and no question of any stigma such as may attach to unemployment.

Allowance for all costs and benefits

In determining what shall be produced and by what methods, the planners can take into account all costs and benefits; without State intervention the market only takes account of private costs and benefits, ignoring externality effects. Where there are social costs and benefits over and above private costs and benefits, whether of production or consumption, central planning, on this score, should result in an output nearer to optimum.

Avoids the wastes of competition

In some ways and some circumstances competition can be wasteful of resources. For example, competitive advertising, aimed not at informing the public but at increasing the market share of each firm at the expense of the others, may mean that each firm uses considerable economic resources simply to cancel out the effects of the other firms' advertising. Such wastes need not occur in a command economy.

Limitations and deficiencies of command economies

Imperfect realisation of claimed merits

In practice the benefits claimed for a command economy are only imperfectly realised or can only be accepted with some qualifications. Thus, take the distribution of incomes. Not everyone will consider that to have wage differentials fixed by one's fellow men is superior to having them determined by the market. Some workers may consider it less invidious if the impersonal market puts a low value on their labour than if a group of planners designate their services of low worth to the community. Again, the merit of central planning, we said, was that it reduced income inequalities. But it is by no means invariably true that earned income differentials are less in a centrally planned economy than under a market system; moreover, whereas in a market economy the same amount of money will always secure the same benefits irrespective of who holds it, in a command economy, with its tight link between political and economic power, there is often inequality of access to material benefits as between, say, members and non-members of the ruling party. Differences in investment income will usually be much less in a command economy because restrictions are usually imposed on the ownership of property and especially on the right of inheritance. However, limitations on inheritance are not, strictly, a distinguishing feature between market and command systems as such. It would be possible to curtail inheritance rights in a market economy without destroying it.

Or take the question of unemployment. While no-one may be designated as unemployed in a command economy, this does not mean that economic hiccoughs do not occur which make workers temporarily idle or cause them to be set to work on pointless tasks because there is nothing better for them to do. Nonetheless, the continuity of wage and the lack of unemployment stigma are important gains.

Again, while a central planning body is able to take all costs and benefits into account, it does not necessarily do so. Quite apart from the difficulties of measuring the externality effect, some public bodies and some countries with centrally planned economies have been prime culprits in the matter of pollution. Nonetheless, the valid point by comparison with a market system is that the latter can only take account of externalities if there is some form of State intervention.

Inefficiency and low living standards

Central economic planning is inseparable from a whole range of inefficiencies. In the absence of prices to serve as an information system and coordinating mechanism, there tends to be both information overload – a massive flow of instructions which is more than operating units can digest – and frequent failures of coordination. A typical example in a centrally planned economy is where newly built flats wait many months for the plumbing or other services to be installed; or flat complexes are completed as housing units while local shopping areas intended for their occupants lag years behind. The coordination failures of the command economy would be even more pronounced were it not for black market activities.

Again, in the absence of factor prices which reflect real costs and relative scarcities, factors of production are not efficiently used. A particularly scarce material may be used in one line of production where substitution could have taken place easily and cheaply, while another product, where substitution was impossible or expensive, is starved of the material. In general there is a lack of incentive and of a competitive spur to efficiency. On the consumers' side there is a loss of welfare because consumers' preferences do not determine production patterns. Although the planners may seek to take account of consumer preferences, they can do so only imperfectly. It is predominantly the planners' preferences which determine the composition of output.

Another aspect of the inefficiency of the command economy is the way in which job vacancies are filled. The market economy provides an automatic efficiency mechanism. Advertised vacancies offering attractive wages will generate more applicants than there are posts, so employers can select those best suited to the job. Anyone vital in his present job would be unlikely to take the post as his present employer would offer him an incentive to stay. Moreover, no-one will apply to whom the move would involve personal hardship. In a command economy, if direction of labour is used without a very careful sifting process, the best man may not be selected and personal hardship may occur. If the planners employ financial incentives, then they are in fact using the methods of the market.

One perennial efficiency problem in a planned economy is the specification of 'success indicators' to enterprises. The planners have to designate output targets for an enterprise. As Lindblom puts it, 'Lacking a common-denominator indicator, any instruction to the enterprise leads it to overplay certain values and neglect others.' (Lindblom 1978, p. 71.) The story may be apocryphal of the nail-making factory whose target output was specified simply in aggregate weight, and which produced one gigantic nail as its annual output, but it points to a very real problem.

If the target is merely for 'tons of nails shorter than two inches' the factory will try to produce all $1\frac{9}{10}''$ nails. If it is for 'numbers of nails' the factory will produce all $\frac{1}{2}''$ nails. But if the plan is set in terms of $\frac{1}{2}''$, $1''$, $1\frac{1}{2}''$ and $1\frac{9}{10}''$ nails, there will be over-centralisation. If the target is set in terms of gross value of output, the factory will maximise its use of materials and semi-fabs and minimise the net value it adds to each product.

(quoted in Lindblom 1978, p. 71)

Another vital deficiency in centrally planned economies is the lack of innovation and technological advance – the system stifles individual enterprise and initiative. On the face of things this assertion may seem incompatible with the progress of Soviet space-rocketry, which, if less than that of the USA, has yet achieved undoubted success. Elst explains the paradox partly on the grounds that 'any large centrally controlled state is capable (at an immense price and to the neglect of everything else) of concentrating all its resources on a few selected strategic centres'. (Elst 1981, p. 29.) However, his main argument is that much of the innovation for military use originates in the civilian sector and that Soviet backwardness has been made good by the importation of Western technology.

The cumulative result of these inefficiencies in the centrally planned economies is a low living standard for its peoples.

Concentration of powers

An outstanding feature of centrally planned economies is the enormous concentration of power in the hands of the State. In a market system a big corporation may distort prices and exercise undue control over people's lives, but this concentration of power is as nothing compared with the complete monopoly of economic and political power by the central planning State. To take one small illustration: when the State employs everyone, if a man falls foul of his employer he can look to no-one else for a job. Correspondingly, if a man falls foul of the State for his political views he can be deprived of all employment.

Philosophical objections

Recognition of this huge concentration of power brings us to the more philosophical objections. To quote Acton's famous dictum:

'All power corrupts and absolute power corrupts absolutely.' Central economic planning is invariably associated with corruption and tyranny. Moreover, when the State controls all the means of production and all output it necessarily controls all the organs of opinion – newspapers, television, radio – and it invariably seeks to control people's minds. Persuasion is one way in which the planners seek to motivate the economy and undoubtedly some people act partly from belief in the ruling ideology. Nevertheless, the underlying sanction is compulsion.

Compulsion as the prime motivating force of the command economy contrasts with incentives in the market economy. Incentives rely on self-interest for their effect and it is this dependence on self-interest which has led to criticism of the moral basis of the market economy. Yet it is not clear that self-interest is any less in evidence in the command economy; it is simply less efficiently harnessed to economic activity and takes a different form. While those at the top of the economic hierarchy act from motives of power-seeking, those lower down act from a combination of hope of advancement and fear.

Capitalism and socialism

In concluding this section it may be helpful to add a note on terminology. The market system is usually associated with capitalism and the centrally planned economy with socialism. Used in this context the terms capitalism and socialism relate to ownership. All advanced economies rest on the existence of capital as we have defined it; the term capitalism implies that the ownership of capital is in private hands, while socialism implies the 'common ownership of the means of production, distribution and exchange'.

We have preferred to avoid these terms because our concern has been with alternative ways of resolving the economic problem, and while in practice the link is close, capitalism is not synonymous with a market system nor central economic planning with a socialist or communist system. Thus the market system is compatible with the existence of nationalised industries or cooperative enterprises if they freely trade in markets; while many of the attributes of the command economy can coexist with private ownership of resources (as in Nazi Germany).

5
THE MIXED ECONOMY: THE UNITED KINGDOM

The mixed economy

It will have become apparent from the previous two chapters that the two extremes of a complete market system with no State economic activity on the one hand and a completely centrally-planned system with no markets on the other hand, simply do not, and indeed cannot, exist in practice within an advanced economy. A market system requires a minimum of legal framework, currency regulations and security from internal disorder or external attack if it is to be able to work and survive. A centrally planned system requires markets for consumer goods and for labour if it is not to revert to barter with all its attendant disadvantages (see Ch. 6). In that sense all economies might be said to be 'mixed'. However, it is more useful to restrict the term 'mixed economy' to one in which there is a major element both of markets and of State economic decision-making; in that sense one is describing a central feature of the economy as compared with, say, the Soviet economy which is predominantly a command economy with peripheral markets, or the British economy at the end of the nineteenth century which was a market economy with some minimal State activity. This concept of a mixed economy does not lend itself to precise definition but it represents the outcome of a more or less conscious decision on the part of a community to reject both the anarchy of unrestricted markets and the apoplexy of central planning; it is an attempt to gain something of merit from each system while avoiding the worst disadvantages of each.

A country may move to a mixed economy from either end of the spectrum. Central economic planning may be modified by the introduction of markets, as happened to some extent in the USSR and more so in Yugoslavia. Or a predominantly *laissez-faire* economy may be modified by an increase in State economic activity. The second form of movement has been the predominant pattern in Western Europe and the USA and characterises the United Kingdom. Henceforward it is the United Kingdom economy on which we shall concentrate.

Forms of State intervention

When a government wishes to intervene in the market system it may do so in one of three broad ways. It may regulate. For example, it could act against polluting producers by designating 'clean air' areas or by forbidding certain forms of water pollution. Or it could protect the public from the external effects of consumption by, for example, requiring motorists to take out third party insurance so that, if the motorist causes an accident, the injured party can be sure of compensation. In both these cases the effect has been to 'internalise' costs which would otherwise have been external. The effect is to improve the working of markets, not to replace them; but not all regulations have this effect.

Secondly, a government may employ financial measures. Thus it might allow producers to continue to pollute but tax the product. If the tax were set at just that level at which the revenue met the cost of offsetting the pollution, then the market system would probably have been improved. (Not all taxes have this beneficial effect, of course. Taxes imposed primarily for revenue purposes are likely to distort markets.) A government could use subsidisation to encourage an activity, like education, because of beneficial external effects. Other forms of financial interventions are transfer payments in cash, met from taxation, to 'improve' the distribution of income – payments to the old, the sick, the unemployed, to those with large families, or simply to those with low incomes. Alternatively, transfer payments may take the form of vouchers, the difference being that vouchers can only be used to pay for a specified good or service; with vouchers the State exercises a control over the direction of expenditure, e.g. food stamps or education vouchers.

Finally, a government can itself provide the good or service. Thus governments provide defence services, police forces and may provide education and health services. Where this happens the market is partly superseded. Thus, in providing education services the government makes the decision about what those services will be and is likely to provide them free of charge (as with infant, junior and secondary education in the United Kingdom) or at a subsidised rate (as, for example, with further education). However, while superseding the market in the supply of the product, the State still utilises markets for the supply of the factors. Thus teachers are hired in the market, the State goes into the market to buy equipment, like books, desks and chalk, and will order new buildings from private contractors on the basis of competitive tender. Even with defence, the usual method in the United Kingdom is to hire the labour for the armed forces in the market (though on the basis of a specially restricted contract) and to buy armaments in the market from private producers. However, sometimes weapons (like atomic bombs) may be made by a government agency, rather than freely purchased in the market and, even if bought from private contractors, special arrangements may be entered into for purposes of security; and in time of war, and even sometimes in time of peace, a government may obtain the labour for the armed forces by direction of labour (conscription) rather than by free markets.

The particular method of government intervention used not only affects whether, or how far, markets are superseded; it also affects the size of the public sector as measured by employment. Regulations, taxation or transfer payments all increase employment in the public sector: regulations have to be administered, taxes collected and transfer payments paid over to beneficiaries. But these methods increase the size of the public sector much less than State provision, when all those supplying the service are taken onto the State's payroll.

Public and private goods

Economists have classified goods according to their suitability for public or market provision. At one extreme are 'pure public goods'. These are the goods or services (discussed in Ch. 3), with diffuse and indivisible benefits, where the exclusion principle cannot be applied and where, under market provision, 'free-riding' could not be prevented. Such services, like defence and the provision of law and order, are eminently suited to public provision and, indeed, it is difficult to have them at all unless the State provides them.

At the other extreme are 'pure private goods', where the benefit is precise and divisible, where the exclusion principle applies and where there are nil or negligible externality effects. A large range of goods and services come into this category, like most food, clothes, biros, gardening equipment and so on. These are eminently suitable for private provision through markets, although this does not mean that a government in a mixed economy may not decide to intervene in the private markets supplying them, e.g. to subsidise basic foods.

In between these extremes lies the area of quasi-public, quasi-private goods. These goods and services *can* be supplied through private markets: there are identifiable benefits and the exclusion principle can be applied. But there are also externality effects which may be considerable. Education and health services come into this category. In a mixed economy like that of the United Kingdom, much of the argument about the relative size of public and private sectors, where the 'social balance' should lie (to use a phrase of Professor J. K. Galbraith (1958)), takes place around these services. The basis of decision is not simply economics; it is also ideology. Even if it could be proved that these services could be more efficiently supplied by private provision through markets there are those who would argue for State provision to ensure a *minimum* standard for all citizens (although this standard could be ensured by vouchers for education and insurance for health). There are also those who argue that the State should not only supply these services but monopolise the supply, to ensure an *equal* standard for all citizens.

In the absence of strong arguments to the contrary (and the forms of market failure outlined in Chapter 3 show that there may be such arguments) there is always a *prima facie* case for provision of goods and services through markets rather than by the State. First, with market provision no-one is forced to buy what he does not want. State provision, on the other hand, save in the unlikely case of unanimity, necessarily involves overriding minorities; there is an irreducible element of compulsion – the difference between a tax and a price. Thus I may strongly object to State expenditure on nuclear arms but I am required, on pain of legal penalty, to make a contribution towards that expenditure. Secondly, where goods are purchased in the market, control rests with those who benefit and there is at least the expectation that benefit will be directly proportional to price. With State-provided goods, however democratic the process, the actual decisions are made by ministers and officials and there is no direct relationship between cost and benefit. Moreover, there are reasons to believe that decisions of ministers are often swayed by considerations of prestige (like building Concorde or putting a man on the moon) or of electoral advantage (like increasing public expenditure in a marginal constituency).

Growth of the public sector in the UK

In the mid-nineteenth century the British economy was very much a *laissez-faire* economy: a market economy with minimal government economic activity. Pure public goods were provided by the State; beyond that, State economic activity was restricted to poor relief, some limited subsidisation of education, and the beginnings of public health provision. The twentieth century has seen a continuously rising trend in public expenditure with fluctuations mainly because of the effect of wars. For all its deficiencies the best single measure of the growth of the public sectors is public expenditure expressed as a percentage of national output; Table 2 shows these figures using representative peacetime years.

The precise figures need to be treated with caution. There is a degree of arbitrariness about just what counts as public expenditure; and somewhat different figures would also be obtained by using the other measures of national output.[1] However, our concern is with

TABLE 2 *UK public expenditure as a percentage of GNP at factor cost (selected years)*

Year	Expenditure as % of GNP	Year	Expenditure as % of GNP
1840	11	1964	38
1890	8	1968	45
1910	12	1972	47
1932	29	1976	52
1951	40	1978	49
1961	38	1980	54

Source: J. Veverka, 1963; *National Income Blue Books*.

trend rather than the absolutes and so what matters most is consistency of definition, which we have sought to attain. The growth trend is very clear from the table.

One or two points need to be made, however, about the meaning of the figure of public expenditure as a percentage of GNP. Take the figure of 54 for 1980; this does not mean that the government pre-empted for its own purposes over half of the national output. Almost exactly half of that 54 per cent represented transfer payments, where government simply transferred money collected in tax to pensioners, students, the unemployed, the sick and other social security beneficiaries and, as interest, to holders of the National Debt. Broadly speaking, the recipients of transfer payments, not the government, determine what the money shall be spent on. Actual expenditure on goods and services by the government was only some 27 per cent of GNP. The point can be put in another way. It will be recalled from Chapter 2 that in arriving at a figure of national income and output we have to deduct transfer payments or we should be doublecounting by taking the same income both in the hands of the taxpayer and the beneficiary of the transfer. To treat the 54 per cent as the proportion of GNP taken by the government would mean including in the numerator what had been specifically excluded from the denominator.

What, then, is the meaning of this percentage? It gives an indication of the size of government spending – the funds passing through government's hands – set against the size of the national output. GNP serves as a datum line, no more. But as such it is useful, enabling us to get away from figures in money terms which, particularly over such a long time-span, are almost meaningless because of changes in the value of money and the size of the economy.

The second point to note is that a global figure of public expenditure can be a misleading indicator of the extent of government economic activity. As mentioned earlier, much depends on the form of intervention. An economy may be very tightly regulated by government with relatively little government expenditure. Further, there may be government trading agencies, including nationalised industries, which do not appear in public expenditure figures, although they represent an important extension of the public sector. The figures in Table 2 relate to central and local government spending and only include nationalised industries and other government trading agencies in so far as the government subsidises them and lends to them.

Reasons for public sector growth

We come now to the $64,000 question: why has the public sector grown so much? We have already implied an answer: to make good the deficiencies of the market system – in particular to temper the outcome of the market system on income distribution and to allow for externalities in services like health and education. However, recognition of market failure is far from being the whole answer. It does not explain the chronology of public sector growth: why the deficiencies were recognised at one time but not earlier. Nor does it explain the extent of public sector growth. Some of the deficiencies of the market could have been made good within the market system by government regulation rather than provision; for example, people might have been required by law to take out health insurance (just as they are required to take out third party car insurance) with possibly the chronic sick looked after directly by the State. Or externalities might have been allowed for by government finance rather than provision; for example, subsidies or vouchers in education. There are clearly other influences at work which we will attempt briefly to explore.

Time-pattern of public expenditure growth In Table 2 peacetime years were chosen to bring out the underlying trend of public expenditure growth; but this has the disadvantage of hiding the pattern of growth. In fact, up to the middle 1950s, a very clear pattern was discernible. In peacetime the level of public expenditure remained fairly steady as a proportion of GNP, but in wartime (and also in the years of the Great Depression, 1929–33) public expenditure rose markedly. After the war it fell back from the wartime peak to stabilise at a new level above the pre-war figure. Over the years public expenditure thus rose in a series of ratchet effects.

Professors Peacock and Wiseman (1967), who discovered this pattern, explain it by the 'displacement effect' of war or other social emergencies. In normal times people accept a certain level of taxation but it is difficult for governments to spend beyond that. But war changes people's ideas of what are acceptable levels of taxation; at the same time it often highlights the need for increased public spending, as when the evacuation of children from city slums in 1939 revealed much malnutrition. War also generates continuing expenditure in the form of pensions to veterans and war widows and interest on an increased National Debt.

The Peacock/Wiseman explanation holds up well for public expenditure growth in the United Kingdom until the mid-1950s. Since then, however, we have had a huge increase in public spending without any wars. It might be argued that the rate of inflation was so high in the period 1973–5, a period of rapid rise in public expenditure, as to constitute a social emergency; but even if this rather doubtful view is accepted, it fails to explain the growth over the rest of the period, especially the period of equally rapid growth, 1964–8. We must clearly look elsewhere for an explanation.

Changes in philosophy Difficult to pinpoint and evaluate, but nonetheless important, were changes in philosophy and attitudes over this period. Socialist and collectivist ideas gained ground with the growth of the Labour Party from early in the century, supported by the extension of the franchise. Then, later, the so-called Keynesian Revolution of the 1940s changed attitudes. The Great Depression had highlighted the evil of mass unemployment and under the influence of Keynesian economics, in the 1944 White Paper, all political parties accepted that to work for a 'high and stable' level of employment was a responsibility of government. The Victorian attitude, that 'money should be left to fructify in the pockets of the people' received a severe blow from the Keynesian analysis that in the absence of adequate investment demand, additional saving by some simply meant less income and employment for others. One way of meeting this situation was to increase public expenditure. The 1950s saw the emergence of a further prophet of public sector expansion, Professor J. K. Galbraith. In *The Affluent Society* (1958) Gabraith argued that public provision would always tend to lag behind private provision. Private goods were advertised but not publicly-provided goods; and emulation – 'Keeping up with the Jones's' – was stronger with private than public goods. As a result, Galbraith argued that to obtain a proper 'social balance' the public sector needed a boost. As so often with prophets, Galbraith probably enjoyed more honour outside his own country than in it.

Change in the nature of publicly-provided goods Growing government expenditure was associated with a change in the nature of publicly-provided goods which changed attitudes to public spending. During the nineteenth century and until well into the twentieth century the majority of public spending was on pure public goods – goods which yielded no clear personal benefit to the individual. In these circumstances it was not surprising that, as Peacock and Wiseman maintained, taxation severely constrained public spending. Taxation was more clearly recognised as a personal burden than public expenditure was recognised as a personal benefit. But as an increasing proportion of public spending began to be devoted to pensions, social security benefits, education and health, so people were conscious of direct benefits which, for many, were more significant than levels of taxation. Demographic trends reinforced the effect. No public expenditure could be of more direct benefit than that on retirement pensions, and the number of people in the appropriate age groups grew steadily throughout the century. The 1911 Census of Population showed 7 per cent of the UK population over the State retirement ages (60 for women and 65 for men); in 1941 the proportion was 12 per cent; and by the 1971 Census, it was 16 per cent.

Producer interests To the consumer benefits of public expenditure must be added the producer interests. The larger the public sector, the more people there are whose wages, salaries and conditions of work are governed by public policy and who have a strong vested interest in increased public spending. Because most of the public sector is concerned with the provision of services rather than goods, wage and salary payments represent the largest part of public expenditure. In recent years public sector workers have become increasingly unionised and the unions have become increasingly militant, so that we have seen strikes by groups such as teachers and civil servants who would formerly have considered such action unprofessional. All this has tended to push up public spending.

Constitutional features A number of what might loosely be termed constitutional features have either added to the pressures for spending, or failed to restrain it. Within government the ministerial heads of the big spending departments make their reputations by new spending programmes rather than by economies. The spending proclivities of the ministers are shared by their civil servants. Moreover, until the recent introduction of cash limits, the one department of the executive concerned with expenditure control, the Treasury, lacked the means to make it effective, especially in times of inflation.

Except for the useful but limited post-mortem work of the Public Accounts Committee, Parliamentary control of public expenditure has been myth rather than reality. Moreover, however much individual MPs may call in general terms for cuts in taxation and reduced public spending, the *specific* pressures they exert are almost all in the opposite direction – like seeking a new hospital or urging the government to aid an ailing firm in their constituency. Moreover, a system of central government grants to local authorities based partly on the size of local spending in the previous year was hardly designed to promote restraint in public spending.

Productivity lag A technical factor has also promoted public spending measured as a percentage of GNP. There is probably much less scope for productivity increases in the public sector, with its predominantly service output, than in the rest of the economy. That being so, if public sector pay keeps pace with the private sector, public expenditure would need to rise simply to enable public sector output to stay the same proportion of GNP in real terms. To put the point in another way: assuming public sector and private sector pay keep in line, then prices will rise faster in the public sector than in the private sector because the bigger increases in private sector productivity will mitigate the price rise there.

Self-sustaining public sector growth Productivity lag, the combined effect of consumer and producer interests in the public sector, the behaviour of ministers, civil servants and MPs, all tend to create a self-sustaining public sector growth which is in fact very difficult to check and more difficult still to reverse. This chapter is being written after two years of Conservative government under Mrs Thatcher pledged to cut public spending. In some services spending has indeed been cut despite great outcries. But it may surprise some readers to know that, overall, in real terms, there has been no reduction in public spending under Mrs Thatcher and, because of the effect of depression in reducing GNP, the denominator in the equation, government spending as a percentage of GNP *rose* by three percentage points between 1979–80 and 1980–1 (see note 1).

In the next two chapters we examine the composition of the private and public sectors of the economy.

6
THE PRIVATE SECTOR IN THE UNITED KINGDOM

Analysis of the private sector

In this chapter we seek to give a synoptic view of the structure of economic activity in the private sector and at the same time describe and explain some of the most important developments and trends. We start with an analysis of business organisations according to type of ownership; we then move on to consider economic activity by 'level' – primary industry, in which we concentrate on agriculture; secondary or manufacturing industry; and tertiary industry, concentrating on retail distribution – and then take a brief look at the financial sector, especially commercial banking, which is relevant to all stages of production. Finally, we look briefly at the size and distribution of the labour force and its organisation in trade unions, which apply to both the private and public sectors of the economy.

Our examination of each stage of production will be very much concerned with size of unit and a word about measurement of size is therefore apposite. Size can be measured in many different ways. Often we have to take what is available rather than the most suitable measure for the purpose in hand. But an appreciation of the characteristics of the different measures improves understanding and helps to avoid errors.

A common measure of the size of a firm is by labour force – number of workers employed, or man/hours or man/years to allow more accurately for part-time labour. For investigations concerning the sources of employment this is obviously the most useful measure, but as a general measure of size it is deficient. The term 'labour' covers workers of many kinds: is it meaningful to regard a firm which employs highly paid and highly skilled workers as equal in size to one which employs only unskilled low paid workers? More significantly, labour is only one factor of production: a firm may be large in terms of number of workers employed but small in terms of capital. The value of capital employed in a firm may be used as a measure of size; but, for companies other than those whose shares are publicly quoted, the value of a firm's capital is difficult to determine and the use of capital alone has similar disadvantages to that of labour. In agriculture the most convenient unit to measure size of farm or estate is often land acreage but this, too, can be very misleading. A market garden, intensively cultivated, may produce a higher value of product than an extensive hill farm, only suitable for sheep grazing, with fifty times its acreage.

In short any measure based on one factor of production – be it labour, capital or land – is deficient as a general measure of size. For this purpose the best measure is value-added – the value of a firm's product less the cost of materials, goods and services which the firm buys in. Value-added is the contribution of a firm to total output; it is equal in value to the payments to the factors of production employed – wages, salaries, profits, interest on money borrowed and rent for hired premises. The snag is that figures of value-added are not generally available.

Another measure which is more often available than value-added is turnover: the value of the product a firm sells without deduction of the value of bought-in materials and services. Turnover will mean different things according to the nature of the product and stage of production. For example, the turnover of a jeweller will be very high in proportion to his value-added because the value of the materials is so high. And a retailer will have a high turnover compared with a wholesaler or manufacturer with the same value-added, because the retailer is at the end of the production chain, buying in goods which already have a high value. But turnover is a useful measure of size for comparing firms in the same line of business and at the same stage of production.

Firms distinguished by type of ownership

Much of the economic theory of the firm is framed in terms of the 'entrepreneur' who manages a business, is the owner of the capital employed in it and who bears the main risks of production.[1] Employees also bear production risks; if the business founders they suffer unemployment. But the entrepreneur stands to lose not only his job but all his belongings. His personal assets as well as his business assets can be taken by creditors if the business fails.

The entrepreneur was a typical figure in the eighteenth-century Industrial Revolution, expanding his business with the aid of 'ploughed back' profits. He might also expand the business by borrowing, on the basis of a contract fixing interest and repayment. The lender in that case only risks the amount lent. The amount an entrepreneur could borrow in this way was fairly closely related to the size of his own assets: the larger his borrowing in proportion to his assets the more the risk that, if the business failed, he would be unable to repay. Thus an individual's scope for acquiring capital was fairly tightly circumscribed.

One method of increasing the capital in the business was by a partnership; partners can be regarded as joint entrepreneurs. Partners jointly own the capital and are jointly responsible for the liabilities of the business. This is a satisfactory arrangement if all partners are active in the business. However, the position of the partner who contributed capital but took no part in running the business was less happy. The so-called 'sleeping partner' received a share of the profits when there were any, but also had as much liability as any other partner to meet obligations to creditors. Nineteenth-century literature contains many stories of the sleeping partner who was rudely awakened by the failure of the business and who was suddenly rendered penniless by his obligation to pay creditors. This actually happened to one famous novelist, Sir Walter Scott, who was a sleeping partner in a publishing firm and found himself saddled with huge debts when his partner absconded. A prolific output enabled him, in the course of time, to meet all his obligations.

In business where large initial capital was required, like canal and railway construction; in industry where machinery was becoming more sophisticated and expensive; and as economies from large-scale production beckoned; so new methods were needed to enable capital to be obtained from a wider circle of contributors who would not incur the risks of the sleeping partner. The answer was the joint stock company with limited liability. Partnership has remained important as a method of business organisation mainly in the professions, where the amount of capital required is very limited and where unlimited liability is felt to be a guarantee of probity – as with solicitors and accountants. But the predominant form of business organisation in the private sector has become the joint stock company.

In theory a joint stock company comes into existence because a group of people jointly decide to take out a share in an enterprise, and they choose the directors who will act for them. Limited liability means that the maximum loss they can incur is the amount they have subscribed (or, in the case of a share not fully paid up, promised to subscribe). In practice, at any rate with a public joint stock company, the company promoters, who may have established a successful business on an individual or partnership basis, or as a private company, appeal to the public to subscribe for shares in what is known as a company flotation. The promoters are then confirmed as directors by the shareholders.

A private company is one in which the number of shareholders is under fifty and shares cannot be freely bought and sold. The private company is not, like the public company, under obligations to publish accounts, the theory being that the shareholders will be in sufficiently close touch to know, or be able to find out, what is going on. Like a public company the shareholders possess the advantage of limited liability. A private company is typically a family business and most small firms are private companies. A few large firms, like C. & J. Clarks the shoemakers, have retained the private company form.

The number of shareholders in a public company is unlimited and shares can be freely bought and sold. The shareholders enjoy limited liability. As a protection for the shareholders various Companies Acts have laid obligations on the directors of public companies. They will incur legal penalties if they raise capital on false pretences, e.g. by mis-statements in a prospectus, and they are bound by certain obligations about publicity, including the requirement to publish annual accounts. Most large firms are public companies.

Many of the largest companies are multinational, having branches in a number of different countries. A multinational company may promote the prosperity of a country, other than its country of origin, by setting up a branch there. But multinational companies are the target of much criticism. There is a natural, if prejudiced, dislike for the intrusion of the foreigner into the national economy. There is concern that a foreign multinational is less under the control of government than a home-based industry, and that it could suddenly generate unemployment by pulling out. There is also the objection that, by means of 'transfer pricing', multinationals may frustrate government policies. Transfer pricing relates to the prices charged for transactions between branches of the multinational. By appropriate transfer prices a multinational company can, for example, arrange for profits to be highest in branches in countries which impose the lowest profits taxes, and lowest in those with the highest profits taxes.

The development of the company form of organisation has brought about a separation between ownership and management. The separation is only partial where company directors are substantial shareholders in the companies they direct, as they often are. But the majority of the shareholders take no active part in the company's affairs. Shareholders elect the directors and dissatisfied shareholders have occasionally acted to remove a board of directors; but usually the election is a formality.[2]

Where companies are managed by salaried directors who may only hold a very small proportion of the total shares, it may be asked: who exercises the entrepreneurial function of risk-taking? The usual answer is the 'ordinary' shareholder. Three main types of publicly issued loan or share capital can be distinguished, with variations. First there is debenture stock. This is a loan stock carrying a fixed rate of interest. Payments of debenture interest represent a cost to the firm and are paid before profits are calculated. Then there are preference shares, which are not very popular these days. Preference shareholders have first claim on the profits but up to a fixed maximum. They bear some of the risk, but not very much; if the firm is making any profits, they will get something. Finally there are the ordinary shareholders (whose shares are also referred to as 'equities'). Ordinary shareholders receive the residue of the distributed profits. They may get nothing, but the sky's the limit. The ordinary shareholder thus bears more risk than any other contributor of capital to the firm, but, because of limited liability and because their jobs are not at stake, ordinary shareholders take less risk than the traditional entrepreneur.

One recent trend has been the growth of the 'institutional investor'. The proportion of personal savings going directly into ordinary shares has fallen in Britain in recent years, not least because of tax reliefs on saving in the form of house purchase, pension schemes and insurance policies. Much of the capital for business now comes from institutional pension funds and insurance companies although, of course, the funds they control are obtained from individuals. Institutional investors tend to be more cautious than personal investors and the effect is probably to inhibit risk-taking and make it more difficult for the smaller, less well-known firms to raise capital on the market.

The forms of business organisation by ownership which we have outlined are the main varieties, but the private sector of the mixed economy offers other possibilities. Worker shareholding is encouraged by some companies and various producer and consumer cooperatives exist. The most famous of these is the retail 'Co-op', which counts all its regular customers as members and of which the distinguishing characteristic, since its inception in the mid-nineteenth century, has been the return of profits to members in the form of a dividend in proportion to purchases.

Agriculture

Primary industry consists of mining, quarrying, fishing and agriculture. In the UK the main mining activity, coal-mining, is nationalised. In this section we concentrate on agriculture.

Agriculture in the UK has a number of interesting features. Perhaps the most surprising is that it only employs some 2.5 per cent of the working population. The number of workers in agriculture has been declining for over a century, reflecting mainly the good record of agriculture in increasing its productivity (measured in terms of output per man). Unlike industry, most of the productivity improvement rests not on machinery but on methods such as better crop rotation, improved stockbreeding, improved strains of plants, chemical fertilisers, more effective pest control, and so on.

The current vital statistics in agriculture have been conveniently summarised by Sutherland (1980): 'The average full-time farm in 1977 had about 230 acres (93 hectares); regularly employed 1.8

workers in addition to the farmer and his wife; had a turnover of £40,000; and required £38,000 of assets other than land and buildings.' Averages can be misleading, but this is less so than most. In England and Wales in the mid-1970s, 25 per cent of land acreage was held in farms within the size range 150–299 acres, with 32 per cent of the land acreage in smaller farms and 42 per cent in larger. At the bottom end, farms below 50 acres accounted for 7.5 per cent of land acreage, while at the top nearly 10 per cent of land was held in farms exceeding 1,000 acres.

Several features of land ownership are of interest. First, the typical farm or estate is individually owned. While some farming companies exist and some land is held by institutions varying from Oxford colleges to insurance companies, the company form of ownership has made comparatively little headway in agriculture.

Further, during the present century there has been something of a silent revolution in land tenure. At the beginning of the century the large majority of agricultural land was held in the form of medium and large tenanted estates; 12–15 per cent of land was in owner-occupation. By the 1970s owner-occupation had become the predominant form of farm tenure, with some 55 per cent of the land owned by its occupiers. The reasons are not wholly clear. In the early stages of the growth of owner-occupation, before and after the First World War, the causal factor seems to have been the break-up of large estates under the influence of depression and of death duties. Tenants bought their farms rather than see them sold over their heads. The continuation of the movement after the Second World War seems to have owed much to rent control, which led to a marked divergence between the vacant possession value of a farm and its tenanted value. This difference in price encouraged landowners to sell if a farm did become vacant; and also encouraged sales between tenants and landlords who split the 'vacant possession premium' between them.

A final point to note is that owner-occupiers who acquired their land some time ago are now one of the richest groups in the community. The average farm of 230 acres which we quoted earlier had an average value in 1977 of £210,000 net of liabilities (which averaged only 10 per cent of net worth). Many farmers bought when prices were low and have benefited from the huge increases in agricultural land prices. In round terms, farm prices rose about 500 per cent between 1945 and 1970 (while retail prices rose by less than 200 per cent) and a further 500 per cent between 1970 and 1979 (while retail prices rose between 200 and 300 per cent).

Manufacturing industry

Table 3 shows the size of units in manufacturing industry measured in terms of employment. A unit for this purpose is an 'establishment', i.e. a factory or plant at a single site or address; a single firm may consist of many individual establishments. The average number of establishments per firm was about 1.33 in 1968.

The table reveals an interesting dichotomy. On the one hand there exist a very large number of small units: over half the units have under ten employees. On the other hand 80 per cent of the workers were in units with more than 100 employees and one-third were in units with over 1,500 employees.

One feature of particular note is the increasing size of plant. Thus, since 1935 the proportion of the manufacturing labour force in firms with over 1,500 employees has more than doubled. Moreover, the growth of plant size since the mid-1930s simply continued an earlier trend.

What is the nature of these economies of scale? First, let us define our terms. By economies of scale we mean that as size increases so costs of production per unit of output (average costs) fall.

TABLE 3 *Size of manufacturing establishments (number of employees) UK 1976*

Size group (employment)	Establishments		Employment	
	No.	%	Thousands	%
1–10	57,000	53.1	267	3.7
11–19	17,000	15.9	246	3.4
20–49	14,400	13.4	445	6.1
50–99	7,500	7.0	527	7.2
100–199	5,000	4.6	694	9.5
200–499	3,900	3.6	1,190	16.3
500–999	1,400	1.3	965	13.2
1,000–1,499	400	0.4	540	7.4
1,500 and over	600	0.6	2,433	33.3
Total	107,200	100.0	7,305	100.0

Source: *Census of Production* 1976.

The reasons for economies of scale are numerous and a full account would take us beyond the scope of the book. They can occur in all aspects of a firm's activities including marketing and finance. We will simply indicate two of the most important sources of scale economies. The first, which applies to all aspects of economic activity, is specialisation, especially specialisation of labour. A larger output may increase the scope for workers, by hand and by brain, to specialise on one or a limited number of tasks. Specialisation enables people to concentrate on those jobs for which they have natural aptitude and inclination and also to acquire skills in that task by training, practice and experience. A modern advanced economy rests essentially on the principle and practice of specialisation. Without it we would be reduced to subsistence economy, with each family unit fending for itself.

A second component of economies of scale applies particularly to manufacturing industry and relates to the phenomenon of 'indivisibility'. Some factors of production, especially sophisticated machinery, come only in a large size. Unless a size of operation is reached which allows indivisible factors to be fully employed, then either they are not used to capacity or an inferior factor has to be used. Some costs, like interest on the money borrowed to pay for the equipment, are the same irrespective of output. Full utilisation of the equipment spreads these costs over more units and reduces costs per unit. Where different indivisible factors are used in conjunction with each other, then the minimum output to enable all the indivisible factors to be used to capacity is the lowest common multiple (LCM) of their outputs. To take a simple example of just two complementary machines. Suppose machine A can turn out thirty parts per week which have to be combined on a one-to-one basis with the output of machine B, which turns out fifty parts per week; then the lowest output to enable both types of machine to be kept running at capacity would be 150 units per week (using five 'A' machines and three 'B' machines).

The scope for economies of scale varies with the nature of the product. Empirical studies suggest that, in many lines of production, economies of scale apply over a particular range of output and then the average cost curve levels off. At some size of output diseconomies will become apparent (i.e. the average cost curve will start to rise). The diseconomies particularly relate to the managerial problems, like coordination and communication, which increase with size, and labour relations are more difficult within very large firms.

The other feature to note about manufacturing industry is its total size. The proportion of output coming from manufacturing in the United Kingdom is low by comparison with similar advanced countries like the USA, France and Germany. Moreover, manufacturing is declining quite rapidly. Between 1955 and 1978 the share of manufacturing in the GDP fell from 37 per cent to 30 per cent, with a somewhat similar fall in numbers employed. In 1979 the manufacturing sector of the British economy employed less than one-third of all employees at work and the slump of the following years must have reduced that proportion still further. Many economists and politicians are concerned at this 'de-industrialisation' of Britain. In 1966 Professor Lord Kaldor was expressing anxiety at what he termed the 'premature maturity' of the British economy. As the rate of productivity increase is generally higher in manufacturing industry than in the rest of the economy, he attributed the comparatively slow rate of Britain's economic growth to this phenomenon. Bacon and Eltis (1976) gave a somewhat different slant to the same theme in arguing that too high a proportion of Britain's labour force was applied to 'non-marketed' goods. The main beneficiaries of the decline in manufacturing employment were the service industries, partly financial services, but especially services provided in the public sector.

Retail distribution

In 1978, as Bamfield (1980) describes in a recent article,[3] 'The 270,000 retail businesses were responsible for a turnover of £45,000m. (of which about £11,700m. was value-added), a labour force of 2.5 million people and 390,000 retail outlets.' Retailing has undergone a major transformation over the past twenty years, comprising in particular the growth of large multiples, a rise in average store size, the development of new forms of retailing such as discounting, the extension of mail sales, and the widespread adoption of new techniques such as self-service. The other side of the coin has been a marked decline in retail outlets and the number of independent retailers.

66 *The economic structure*

TABLE 4 *Share of retail trade 1961–78 (%)*

	1961	1966	1971	1978
Cooperatives	10.9	9.1	7.1	6.8
Multiples	29.2	34.5	38.5	46.5
Independents	59.9	56.4	54.4	46.7

Source: Bamfield 1980.

Tables 4 and 5 show the changes in the share of the retail trade by different categories of retailer and also the decline in numbers of shops. The definitions used in the tables are as follows. 'Cooperatives' are retail businesses owned by consumers (varying in turnover from under £50,000 to over £400,000); 'multiples' are retail groups owning ten or more stores (with the biggest owning hundreds of shops); 'independents' are firms with less than ten shops; 'large independents' are those with two to nine shops.

Between 1961 and 1978 the multiples expanded their share of retail trade by some 60 per cent at the expense both of the cooperatives and the independents. All forms of business have registered a decline in numbers of shops. The multiples and especially *the* Co-op have cut out unprofitable outlets; but in absolute terms the big decline has been in one-shop independents which fell, between 1971 and 1978, by 93,000. This was a decline in one-shop independents of 28 per cent, but accounted for over three-quarters of the total decline in number of shops.

The main element in this transformation has been the dynamic of the multiples, but the small shops also suffered some particular disadvantages.

The number of supermarkets (defined as a self-service food store

TABLE 5 *Shop numbers 1971–8*

	1971	1978	% change	Sales per shop 1978
Cooperatives	16,480	10,370	− 37.1	£295,950
Multiples	71,162	66,343	− 6.8	£315,165
Large independents	83,966	67,886	− 19.1	£95,911
One-shop independents	338,210	245,000	− 27.6	£59,105
All traders	509,818	389,599	− 23.6	£115,426

Source: Bamfield 1980.

with a sales area in excess of 2,000 square feet) increased from eighty in 1957 to 2,700 in 1966 and 6,200 in 1978, the average size of supermarkets rising continually over the period; multiple firms exploited the benefit of supermarkets to the full, controlling over 50 per cent of the grocery trade by 1978. The formula for their success was price-cutting, no credit, careful selection and presentation of goods and self-service methods. Multiples were also able to take advantage of inner-city developments since the early 1960s to acquire many prime sites. In other lines than food, multiples have gone in for diversification. Much enterprise has been shown by multiple firms in introducing new and sophisticated techniques, such as computer-linked regional distribution centres. Within the multiple firm sector, partly as a result of takeovers, there has been an increasing concentration of trade in the hands of a few very large firms.

Small shops have always tended to be less efficient than large. Their labour costs tend to be higher because of the 'indivisibility' problem – if the shop was to remain open at least one person had always to be present even if he was kept far from fully employed. This relative disadvantage was increased by the labour savings effected in many of the multiples by self-service. In addition, certain increases in operating costs, partly indirect effects of government policy, have hit the small shops. Their fuel costs tend to be disproportionately high and fuel has risen disproportionately in price. They have suffered from the rise in rates – out-dated rating valuation lists have favoured their better-placed rivals (see Ch. 7). It is also no coincidence that the decline in numbers of shops accelerated from the early 1970s, when VAT was introduced, which placed compliance burdens on traders which were especially onerous for small firms (Sandford *et al.* 1981).

Financial services – the banks

The range of institutions providing financial services in the United Kingdom is wide, immensely varied and infinitely complex in its inter-relationships. Central to this financial activity is 'The City'. The City is in one sense a city within a city, the square mile in the oldest part of London in which are concentrated the offices of banks, discount houses, insurance companies, shipping agencies,

68 *The economic structure*

together with the Bank of England, the Stock Exchange and commodity markets. But as William Clarke puts it in his excellent book, '"The City" is in effect used as a convenient shorthand for the hub of all the financial and commercial activities that go on in the UK, for a score of different markets and industries rather than geographically to describe a place on a map.' (Clarke, *Inside the City* 1979, p. 3.)

In our synoptic view of the private sector of the economy we must restrict ourselves to a compressed analysis and a few brief comments on the banking system. But, to provide a perspective, it is worth quoting Clarke's summary list of the basic roles which the City plays.

It provides capital and services for industry and government.
It channels and invests the country's savings.
It operates leading domestic markets in commodities, securities and money.
It operates *world* markets in Eurocurrencies, foreign exchange, gold, insurance, shipping and commodities.
It provides advisory services in accountancy, investment, law and commerce.
It attracts and communicates financial and economic information to and from the rest of the world.
It earns valuable foreign exchange.

Even to say we shall restrict ourselves to the 'banking system' is something of a misnomer. Apart from the National Savings Bank, Trustee Savings Banks, building societies and many other institutions which might reasonably be regarded as part of the banking *system*, there are four different categories of bank: clearing banks, the merchant banks offering specialist banking services, the British overseas banks (with head offices in London and branches all over the world) and the foreign banks, of which there are well over 300 in the City, almost all of which have moved in during the past twenty years. In fact, beyond acknowledging the presence of these other banks we shall confine our attention to the clearing banks.

The term 'clearing banks' describes the six banks which are members of the Committee of London Clearing Bankers; they jointly own and control the Bankers' Clearing House which administers and runs the clearing of payments within the British banking system. Before 1968 there were eleven clearing banks, but mergers in 1968 and 1970 reduced them to six and they are dominated by the 'big four': Barclays, National Westminster,

Lloyds and Midland. Between them they have something like 12,000 branches in England and Wales. Three Scottish banks do similar business in Scotland with 1,500 branches and there are four banks in Northern Ireland with about 400 branches.

The prime function of the clearing banks is to provide the facilities for transferring money from one person or company to another.[4] At one time the forerunners of the present clearing banks used to issue their own bank notes as money, backed by gold.[5] The terms of the Bank Charter Act of 1844 gradually resulted in the concentration of note issue in the hands of the Central Bank, the Bank of England (see Ch. 7). However, the clearing banks, as well as using Bank of England notes, provide their own form of bank money, i.e. bank deposits transferable by cheque.

Intimately connected with the process of transferring money is the borrowing and lending of money by the clearing banks. They borrow, through their branches, both surplus balances on current accounts not usually bearing interest and on interest-bearing deposit or time-deposit accounts repayable at seven days' notice. The traditional form of lending has been by overdraft. An agreed facility is granted to a customer to draw on a current account, and interest is only paid on the amount actually borrowed. On the other hand the overdraft is strictly repayable on demand, though usually the facility can be extended. Because the borrowing is short-term, traditionally the banks have only lent on a short-term basis, e.g. to finance firms in the purchase of materials which would, within a limited time period, be turned into finished goods and sold. Business was expected to look to the Stock Exchange for long-term capital. As insurance against depositors requesting cash for their deposits, banks have always kept a proportion of their assets in liquid form (e.g. Treasury Bills, a form of short-term borrowing by the government which can be turned into cash at short notice with minimum risk of loss).

Twenty years ago the clearing banks were regarded as rather cosy, conservative institutions engaging in little competition with each other. Since then, and especially since a change in government policy in 1971 known as competition and credit control, the clearing banks have become far more innovative. Increasing competition from building societies for deposits and the influx of foreign banks has been a further stimulus.

On the payments side they have developed credit cards (led by Barclays) and cash dispensers. They have reorganised their lending to industry, introducing 'term loans' of between, say, five and seven years, and enormously expanded this form of lending. Following the Midland's lead they have developed a system of personal loans with built-in insurance. They have also began to move into the area of business traditionally the field of the merchant banks, helping with new issues of shares, giving financial advice to companies and assisting with export finance. These moves have been aided by some mergers of merchant banks with clearing banks. In retaliation to the competition of US banks they have also set up their own branches in the main financial centres of the world.

The City generally and the banks in particular have been criticised in recent years on the grounds that they were not offering enough assistance to industry. On this score the committee under Sir Harold Wilson which examined the City institutions gave them a comparatively clean bill of health.

Trade unions and the labour force

In September 1979 the working population in the UK was estimated at about 26.5 million. The term 'working population' is defined to include both those in and those seeking work, and this total comprised just under 23 million employees, some 1,900,000 self-employed, 320,000 in HM Forces and 1,400,000 unemployed.

The number of male employees has fallen somewhat since a peak in 1965. More significantly, the number of females at work has increased by about a quarter since 1960. In 1977 as many as 40 per cent of these females worked part-time (less than thirty hours per week) and the majority of part-time workers were wives. An outstanding economic fact of postwar Britain has been the rise in the married female 'activity' or 'participation' rates (the proportion of an age/sex group in the working population). The activity rate for all wives increased from 22 per cent in 1951 to 49 per cent in 1976, while that of the age group 45–59 rose from 22 to 61 per cent over the same period. While the reasons must partly be sought in demand changes in the labour market, the big increase in working wives undoubtedly reflects changes in social attitudes. It is probably also, in part, an outcome of the increased availability of consumer durables like vacuum cleaners, washing machines, fridges, dishwashers and cars, which make it easier for married couples to combine housekeeping with paid work for both of them. The supermarkets, described in the previous section of this chapter, cater very much for this type of mobile working couple with ready cash.

We have already touched on some of the changes in the distribution of the labour force, notably the decline in employment in manufacturing industry and the growth in the service sector. In the mid-1950s the manufacturing and service sectors each employed about 42.5 per cent of the labour force. By the beginning of 1980 the manufacturing sector was down to 31 per cent and the service sector accounted for well over half total employment. Particular growth areas have been professional and scientific services, especially education and medical and dental services; but it should be noted that a high proportion of the women in these services work only part-time.

Rather more than one-half of the labour force is organised in trade unions following a rapid period of trade union growth in the 1970s. Male membership is higher than female membership, but female membership has been rising the more rapidly. Manual workers, semi-skilled workers and workers in the manufacturing and public sectors are among the most highly unionised.

The number of unions has been falling for many years but, in 1980, was still over 400 with over 50 having less than 100 members.

In their monumental study on the history of trade unionism, Sidney and Beatrice Webb defined a trade union as 'a continuous association of wage earners for the purpose of maintaining or improving the conditions of their employment' (*The History of Trade Unionism*, 2nd edn, 1896, p. 1). This is a useful starting point; but 'wage earners' must not be interpreted narrowly, for unionisation is widespread among salaried workers in the public sector and elsewhere, e.g. in banking. Further, some of the earliest activities of trade unions, and more especially some of the more recent, go beyond or at any rate place an exceptionally broad interpretation on the phrase 'conditions of their employment'.

The essential function of trade unions has been to overcome the weakness of the bargaining power of individual workers by substituting collective bargaining for individual bargaining, thereby raising the supply price at which labour is sold and making the rate of wages uniform over a whole trade. Collective bargaining implies the use of the strike if the bargainers fail to come to terms. The demand for a control of working conditions in the factory, as well as interpretation of piece-rate scales and the like, has developed as a natural corollary of bargaining on pay and hours of work; to bargain on pay alone would be futile if there was not some check on working conditions.

Historically many unions have had a very strong welfare or Friendly Society side, with provisions for sick pay, unemployment benefit, death grants, pensions and sometimes emigration funds. The growth of State provision during the twentieth century has reduced the importance of this aspect of union activity.

The trade union movement provided the basis for the formation of the Labour Party early this century and unions have a formal place in the Labour Party constitution which gives them a dominating voice in the Annual Conference decisions of the Party and in its National Executive Committee. The Labour Party also gets most of its funds from the unions. About a third of Labour MPs are sponsored by trade unions. Unions are, however, required by law to keep their political funds separate from their other funds and union members have the right to contract out of the political levy. Trade unionists, often nominees of the Trades Union Congress (TUC), serve on most committees or commissions set up by government on social and economic matters, including the National Economic Development Council.

The activities of trade unions raise some fundamental issues of politics, sociology and law as well as economics. We can do no more here than indicate some of the main areas of concern and controversy. For a number of reasons trade union power has grown in recent years. Advanced technology has made employers more vulnerable to strikes and sometimes a selective strike by a few key workers can halt a whole works or cause considerable hardship to the community. The payment of income tax rebates to strikers and supplementary benefit (untaxed) to strikers' families has reduced hardship to strikers and economised union funds. Unions also gained power under the legislation, 1974-6, of the Labour Government.

Public opinion polls have indicated that the general public shares the view of many politicians and commentators that the unions have become too powerful and their powers need to be curtailed, but attempts to do so have achieved little. In 1969-70 the Labour Government abandoned its proposed reforms in the face of strong union and back-bench opposition. The Conservatives pushed through an Industrial Relations Act in 1971, but it aroused bitter union hostility and was repealed by the Labour Government of 1974 which went on to strengthen the position of trade unions (e.g. it ceased to be unfair practice for an employer to dismiss an employee for refusing to join a union when there was a closed shop agreement (100 per cent union membership) between employer and unions). The Conservative Government of 1979 came to power committed to trade union reform, but in the light of the 1971 experience proceeded on a softly, softly basis.

Perhaps the main questions, which are far from being settled, are (1) What degree of legal immunity should unions possess? For example, they cannot be sued for damage committed by their members in the course of industrial action. (2) Should workers in activities crucial to the community, such as firemen, police and civil servants, be denied the right to strike in exchange for some guarantees about wage levels? (3) Under what conditions, if at all, should a man be sacked for refusing to join a union (the closed shop issue)? (4) What are the proper limits to picketing? Ought there to be restrictions on numbers of pickets to prevent intimidation? What constraints should be imposed on place of picketing? Should picketing of suppliers' premises be illegal? (5) What action, if any, should be taken to promote union democracy, to see that union officers genuinely reflect the views of members? For example, should there be a compulsory secret ballot in elections for union officers and on decisions involving industrial action?

Besides these questions, which might be capable of resolution by changes in trade union law or by a code of practice, there remain broader issues of economic and political significance. One is the problem of how to reconcile free collective bargaining with the control of inflation. A second concerns the proper limits of trade union political action. Some recent strikes, most notably the miners' strike which brought down the Heath Government of 1974, have had strong political overtones. Further, is there some limit beyond which the close association between Labour Party and unions threatens genuine political democracy? The TUC often appeared to be exercising a dominant influence over the Labour Government of 1974; and, unless changed by an amending decision of Conference, the trade union block vote will in future be the largest single influence in determining who shall be Leader of the Labour Party, and hence the next Labour Prime Minister.

7

THE PUBLIC SECTOR IN THE UNITED KINGDOM

Scope of the public sector

In this chapter we look at the scope and composition of the public sector. In the first section we examine the composition of total public spending, give some indications of how it has changed and is changing, and then look briefly at the taxes which finance it. In the second section we concentrate on local government spending, how that is financed and the problems associated with the one local tax, the local rate. Section three surveys the nationalised industries, with a particular comment on the Bank of England. In the final section we touch on some of the problems of administering a huge governmental machine and consider the economics of bureaucracy.

We should repeat the warning that a description of the size and main directions of government spending, even including the nationalised industries, does not indicate the full extent of State control of, and influence on, the economy. Regulations on planning and pollution, measures to promote competition and restrict monopoly, and changes in taxation and monetary control, may have a major effect on the economy for a negligible direct effect on public spending.

The composition of public spending

The main headings of government spending, central and local combined, are set out in Table 6. A few words of explanation are required. The table is derived from the most recent of the Public Expenditure White Papers available at the time of writing. Each year a committee of civil servants, known as the Public Expenditure

TABLE 6 *Main components of government spending, UK (measured at 1980 survey prices)*

	Actual 1975–6		Estimated 1980–1		Planned 1983–4	
	(£m.)	(%)	(£m.)	(%)	(£m.)	(%)
Defence	9,436	11.6	9,746	12.3	10,350	13.6
Overseas aid/other overseas services	1,176	1.4	1,648	2.1	1,530	2.00
Agriculture, fisheries, food	2,432	3.0	1,182	1.5	980	1.3
Industry, energy, trade, employment	5,136	6.3	4,386	5.5	2,900	3.8
Government lending to nationalised industries	1,694	2.1	2,050	2.6	−150	−0.2
Transport	4,796	5.9	3,528	4.5	3,290	4.3
Housing	7,414	9.1	5,361	6.8	2,980	3.9
Other environmental services	4,764	5.9	4,008	5.1	3,670	4.8
Law, order and protective services	3,087	3.8	3,371	4.3	3,530	4.6
Education, science, arts, libraries	11,376	14.0	10,747	13.6	9,920	13.1
Health and personal social services	10,654	13.1	11,263	14.2	11,800	15.5
Social security	16,203	19.9	20,390	25.7	22,090	29.1
Miscellaneous and adjustments*	3,116	3.8	1,564	2.0	3,115	4.1
Planning total	81,283	100.0	79,245	100.0	76,000	100.0

* Includes contingency allowance of £193 for 1980–1 and £1,500 net of shortfall for 1983–4.

Source: *The Government's Expenditure Plans 1981–82 to 1983–84*, Cmnd 8175, March 1981, Table 4.1.

Survey Committee (PESC), collates the expenditure estimates of each department for the current year and their forecasts of expenditure for several years ahead. These forecasts are then modified by ministers in the light of government policy and agreed figures arrived at to give public expenditure 'planning totals' for the next few years. In order that comparison can readily be made one year with another, the expenditure for all the years is made in terms of the prices ruling at the time of the survey. In the table we have shown three years, 1975–6, 1980–1 and 1983–4, to give a perspective about the actual and expected direction of change in the expenditure programmes. The Public Expenditure White Paper was published just before the end of the 1980–1 financial year, and therefore the 1980-1 figures are only estimates, but they are reliable estimates. The 1983-4 figures of planned expenditure are much less reliable; governments have not been very good at sticking to their expenditure plans. In considering the year 1983-4 it should be noted that there is a substantial contingency sum included in the 'miscellaneous and adjustments' line; this sum is likely to be added to the individual totals when the money comes to be spent in 1983-4. One item of expenditure excluded from the government's planning totals of expenditure, on the grounds that it is not susceptible to planning decisions, is debt interest. In 1980-1 the *gross* figure of debt interest (mainly offset by income from assets against which the debt had been incurred) was expected to be £10,820m., or 13.7 per cent of the planning total for that year.

Taking the long view and comparing present expenditure with the position a century or more ago, the big increase has been expenditure on the social services. Education, health and personal social services, and social security, together amount to well over half (53.5 per cent) of expenditure, 1980-1. If housing is added, the total rises to over 60 per cent. By comparison expenditure on the social services amounted to only 9 per cent of total government spending (including debt interest) in 1840 and 20 per cent in 1890. In these periods expenditure was defence-dominated. The change reflects a dual transformation. First, a change in approach to poverty, which seeks not simply to relieve it (often a niggardly and harsh way) but rather to cure it by dealing with its causes, like old age, sickness, unemployment and large families. Second, a change from services designed simply to deal with poverty, to welfare services for all citizens.

If we look at more recent changes and at plans for the next few years, defence and expenditure on law and order are both on a modest upward trend; support for industry both by way of general subsidies and of loans to nationalised industries is expected to fall markedly. Housing expenditure is on a pronounced downward trend. Education is falling, while health and personal social services are rising, each reflecting different demographic trends – the decline in numbers of the young and the increase in numbers of the old. The big increase in social security payments since 1975-6 has three main components. First, a substantial rise in expenditure on retirement pensions and on supplementary pensions, reflecting the growing numbers over retirement age. These two items amounted to just over half the social security payments in 1980-1 and the upward trend here is due to continue. Second, a big increase in unemployment benefit and in supplementary benefit for the unemployed consequent on the slump; expenditure on this head is expected to have fallen by 1983-4, but not by very much. Third, a big jump in expenditure on child benefit in the late 1970s, as child tax allowances were replaced by a new scheme of child credits.

The finance of expenditure

Some £13 billion of public expenditure in 1980-1 was financed by borrowing. The bulk of the remainder came from central government taxation, national insurance contributions and local rates. National insurance contributions, as a compulsory levy for most payers, can reasonably be treated as a tax. On this basis Table 7 shows the contributions of the main taxes to government revenue on the basis of the provisional out-turn, 1980-1.

Some 60 per cent of tax revenue comes from taxes on income, profits and capital, and 40 per cent from taxes on goods and services (including local rates which are a tax on the use of buildings).

One feature of the income tax, which contributes some 30 per cent of tax revenue, is unusual if not unique. A so-called 'basic' rate

TABLE 7 *Contribution of main taxes to UK revenue 1980-1 (%)*

Income taxes	29.6
National Insurance contributions	17.2
Value-added tax	13.5
Local rates	10.4
Corporation tax	5.6
National Insurance surcharge	4.3
Oil	4.2
Tobacco	3.3
Alcoholic drinks	3.1
Petroleum revenue tax	2.9
Remainder	5.9
	100.0

Note: Taxes are indicated individually where receipts exceed 2 per cent of tax revenue. The percentages are based on the figures of provisional out-turn in *Financial Statement and Budget Report, 1981-82*, HMSO, March 1981.

of tax (standing at 30 per cent in 1980-1) applies to a very wide band of income, so wide in fact that upwards of 90 per cent of taxpayers are subject to this rate. As a result, the progressiveness of the income tax is very limited except for those whose incomes bring them well into the higher rate band. Indeed, taken overall, the progression of the UK tax structure is very modest. Most redistribution of income effected through the State comes about through the receipt of benefits rather than the payment of taxes.

Local government expenditure and revenue

Local government in Britain today supplies an impressive range of services, from those affecting everyone, like police and fire services, to those which cater for more specialist tastes like museums and art galleries. The predominant item of expenditure is education, which takes half or more of the budget for current expenditure. Local government employment grew markedly in the 1960s and 1970s; between 1961 and 1978 it rose by just over 60 per cent. This increase was mainly caused by growth in the education sector; but that sector is now being cut back.

By any measurement local government is big business. As the Report of the Committee of Inquiry into Local Government Finance (Layfield Committee 1976) put it:

The scale of provision has reached great proportions. Today's local government spends some £13,000m. a year on its work. It services a capital debt of some £25,000m. Nearly 3 million people are now employed in the service of local authorities. These authorities are also among the biggest landlords and landowners in the country. Together they comprise one of the biggest enterprises in the land.

The figures in the quotation relate to 1975-6. The level of estimated expenditure in 1980-1 is set out in Table 8, which is on the same basis and from the same source as the data on total government spending in Table 6, except that the figures exclude Northern Ireland.

Some of the most difficult problems, with both an economic and a political dimension, relate to the finance of local government. Apart from income from fees charged for services (like car parking and council house rents) about 60 per cent of local authority

TABLE 8 *Estimated public expenditure by local authorities, Great Britain, 1980–1 (at 1980 survey prices)*

	Current expenditure (£m.)	Capital expenditure (£m.)	Total (£m.)	(%)
Education and science, arts and libraries	8,286	414	8,700	44.3
Local environmental services	1,841	533	2,374	12.1
Law, order and protective services	2,273	76	2,349	12.0
Personal social services	1,447	86	1,533	7.8
Transport	1,281	704	1,985	10.1
Housing	622	1,516	2,138	10.9
Other programmes	384	164	548	2.8
Totals	16,132	3,492	19,624	100.0

Source: *The Government's Expenditure Plans 1981–82 to 1983–84*, Cmnd 8175, March 1981, Table 1.5.

income (in aggregate) comes from central government grants and 40 per cent from local rates. Rather more than half of rate income comes from non-domestic rates. The position is illustrated in the simple block diagram, Fig. 2.

The majority of grant income, before 1980, came from a Rate Support Grant, which attempted to take account of the 'needs' and 'resources' of individual authorities and also, from 1967, to give an element of relief to domestic ratepayers. The system was extremely complicated and imperfect; moreover, past levels of expenditure were taken as one of the indications of need. Thus high spending generated more grant. Mrs Thatcher's Conservative Government has attempted to change this situation by making its own assessment of the appropriate expenditure levels of individual local

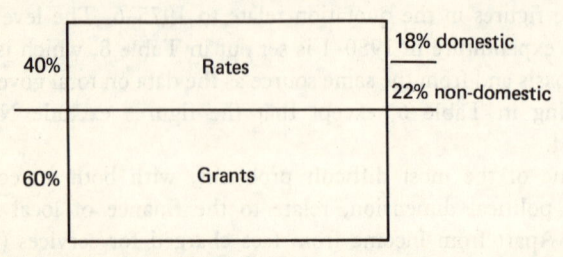

Fig. 2 Sources of local government finance

authorities and penalising overspenders. But this is a detailed, difficult and somewhat arbitrary exercise.

Three main issues are raised by central government grant support to local authorities. How much grant is compatible with genuine local autonomy? How do you allocate that grant without encouraging irresponsible local spending? How can you reconcile the needs of economic policy, as seen by central government, with local autonomy?

The complementary issue to the size and nature of government grant is the form and extent of an independent source of income for local authorities. Apart from charges for services, their one such source is the local rate. Rates are a tax charged at so much in the pound of an assessed annual value (the rateable value) of real property (mainly land and buildings) and payable by the occupier. Of the non-domestic rates, rather more than 80 per cent of the revenue comes from commercial and industrial property.

Domestic rates have come in for much public criticism. The Layfield Committee was set up in response to an outcry against them and the Conservative Party is on record with the promise to abolish domestic rates – though implementation has been indefinitely postponed. The main criticism has been that domestic rates are regressive (constitute a larger proportion of the income of poorer households than of better-off households); that they are a tax on a necessity, shelter; and that they are inequitable because many people who benefit from local government services are not householders and therefore pay no rates. These arguments are not without force. However, the worst effects of the regressiveness of rates have been eased by rate rebates for the poorest households. Moreover, rates do have particular advantages as a local tax: housing is widely dispersed throughout the country and so rates generate revenue in all local areas; rates are economical to administer on a small scale; the tax base is visible and fixed within a local authority area; and rates are clearly perceptible as a local tax, and hence promote local accountability.

The case against rates on industrial and commercial property is much stronger than that against domestic rates. Business rates tend to reduce investment and distort investment and location patterns. They are a particular burden on new firms (which have not yet reached the profitable stage) and on firms going through a slump, because, unlike income or profits taxes, rates have to be paid even though no profits are made. In so far as they are passed forward into prices (which is more difficult in a slump) the evidence suggests that they are regressive (with no offsetting rebate). The lack of regular revaluations has caused inequities and in particular has hit small firms in run-down inner city areas, who ought to be paying less rates compared with new firms in prime sites. Above all, however, non-domestic rates do nothing to promote 'accountability'.

The principle of accountability in local government was the keynote of the Layfield Report. Layfield stressed the need to establish 'a financial system based on a clear identification of responsibility for expenditure and for the taxation to finance it'; and argued that: 'Whoever is responsible for spending money should also be responsible for raising it so that the amount of expenditure is subject to democratic control.' In so far as local income comes from domestic rates there is democratic control; but no such control results from rates on business property. As we have seen (Fig. 2) only 18 per cent of local government income, in aggregate, comes from domestic rates. In some authorities the proportion is much less because they receive larger grants, and/or they contain a high proportion of non-domestic property, and/or a relatively high proportion of households receive rate rebates. In these circumstances the link between spending and raising is weak indeed. Such councils can raise rates without fear of electoral penalty and at the expense of the commercial and industrial prosperity of the area.

Layfield recommended the retention of both domestic and non-domestic rates, though it came near to advocating the abolition of the latter (Jones 1978). It proposed, however, that the load placed on the rates should be reduced and that the larger authorities should raise money by a local income tax (as in many continental countries). Both Labour and Conservatives have shown no enthusiasm for this proposal. In the absence of some additional form of local revenue, the democratic foundation of local government must remain very shaky indeed.

There is also a case for a radical review of the charging policy of local authorities. Many services for which fees are charged are priced below their economic cost with no obvious social justification. Why, for example, should the rate-paying widow of modest means subsidise the comfortably-off young athlete at the local sports centre? Or why should the pedestrian, who can't afford a car, subsidise the parking facilities for motorists? The justification is also dubious for the biggest area of local authority subsidisation, council house rents. (It is as dubious as the reliefs given to owner-occupiers through the income tax system.) Where people of different income levels live in council houses, the general subsidisation of council house rents is a very imperfect way of reducing income inequalities. Finally, to make an economic charge to industry and commerce for services which specifically benefit them, like refuse collection, might be a more appropriate way of securing their contribution to local revenue than the non-domestic rate.

The nationalised industries

The trading bodies which have been in public ownership for a considerable time fall into three broad groups: fuel and power industries – gas, electricity and coal; industries providing transport and communication services – British Rail, National and Scottish Bus Company, National Freight Corporation, British Airways and the Post Office (covering Telecommunications and Postal Services, now separated); and two manufacturing businesses – British Steel Corporation and British Leyland.

These bodies were taken into public ownership for a variety of reasons: some were natural monopolies (like the postal service and British Rail); others were taken over because of a history of bad industrial relations (like coal-mining); in other cases nationalisation was to save them from bankruptcy (like British Leyland). The nationalising governments were generally inspired by socialist ideology: that nationalisation of 'basic' industries would give the State a greater command over the economy; if they were owned by 'the nation' rather than by capitalists seeking private profit, industrial strife would necessarily be reduced; eliminating competition would reduce waste; and in public ownership they could be run not only more fairly for their employees but also more efficiently.

Richard Pryke (1981) has conveniently summarised the importance in the economy of these eleven publicly-owned industries. They are jointly responsible for nearly 10 per cent of total output. 'They range in size from Telecommunications which, during 1977, produced almost 2 per cent of GDP, to the Freight Corporation, which accounted for only 0.2 per cent.' Electricity, gas, coal and British Rail all contribute 1 per cent or more to GDP, while postal services, British Steel Corporation and British Leyland each produce between 0.5 and 1 per cent. To quote Pryke again: 'In 1977 they had about 1,750,000 workers, which represented 7 per cent of total employment and they were responsible for over 16 per cent of gross fixed capital formation excluding housing.' (Pryke 1981, p. 1.)

At the risk of oversimplification by generalising across the whole field, let us try to assess how they have fared and how far the hopes for them have been fulfilled. The ideal comparison is, of course, the hypothetical one of their performance if they had never been nationalised with their actual performance. That we shall never know. The best we can do is to compare them with private industry in this country, especially any industry or firms in a similar line of business, and to compare them with similar industries abroad.

In recent years some of the bitterest and most prolonged industrial conflicts have been in the nationalised industries, notably the miners' strikes of 1972 and 1974 and the steel strike of 1980. Nationalisation has introduced into industrial relations an aspect not appreciated at the time of nationalisation. Governments have a special responsibility for a nationalised industry. That being so, the market does not impose the same restraints on wage demands in nationalised industries as in private industry, for the government can always be seen as the source of funds to meet a wage demand. Further, in that situation, a strike to some extent is necessarily politicised. It is very difficult for the government to stand aloof. A strike thus becomes a strike against the government.

Not only has strife not been absent from industrial relations in the nationalised industries but the job satisfaction of the workers appears to have been low. The results of a survey of 24,000 workers, by questionnaire, published in *Money Which?* (1977) show workers in nationalised industries as those least satisfied with their jobs – as compared with the self-employed, government employees and even employees in private industry.

On efficiency their record, too, has been extremely disappointing. Richard Pryke (1981) has examined their performance since 1968 and compared them with the most appropriate alternatives. He concludes: 'The performance of the nationalised industries over the past decade has ranged from being good in parts – telecommunications and gas – to being almost wholly bad – BSC and postal services.' Most of the industries have displayed serious inefficiency in their use of labour and capital and misdirected resources because of widespread failure to pursue optimum policies for pricing and production. Pryke's verdict is the more impressive because formerly he was an advocate of nationalisation.

Much of the poor performance arises from arbitrary interventions by governments which have used the nationalised industries in pursuit of national economic policies at the cost of undermining the financial guidelines and demoralising the industries. But not all the failures can be excused in this way.

There remains a need for more experimentation in the form and structure of nationalised industries. In some industries there is scope for more competition. In others, where there is a natural monopoly, a move away from a unitary form of organisation may be desirable. And what more suitable area than nationalised industries to experiment with worker participation on the board of directors?

The Bank of England

A rather special nationalised industry is the Bank of England – and no book on economic structure would be complete without some reference to the central bank. One of the early acts of the Labour Government of 1945 was to nationalise 'the Bank'. In fact, however, nationalisation made little difference. Whether nationalised or not, the prime function of a central bank is to carry out the monetary policy of the State. In pursuit of this objective the Bank of England has certain powers and responsibilities. It has a monopoly of the note issue in England and Wales. It acts as the government's bank, keeping the government's balances, managing the National Debt, evening out temporary shortages of money, lending to the government as required and acting as the government's broker in placing Treasury bills and government bonds. It also intervenes in the foreign exchanges, buying or selling sterling in exchange for foreign currencies, in accordance with government exchange rate policy. The Bank of England, too, is the bankers' bank. The London clearing banks and a few other banks hold balances with the Bank of England and can adjust inter-bank indebtedness by means of these balances. Through these balances the Bank of England can operate on the total volume of money. The Bank of England may also act as lender of last resort. If the commercial banks call in their short-term loans, the Bank of England will replace them with its own loans – but at a price. It also might take the initiative in organising a rescue operation if a bank got into difficulties, in order to maintain the stability of the monetary system.

The economics of bureaucracy

In this final section we look at some of the problems that particularly arise with a large public sector. We have headed the section 'The economics of bureaucracy' but the first kind of problem we examine is one which arises with any large-scale organisation rather than specifically with bureaucracy as we define it below. However, 'the State' has come to be the largest organisation in the community and is therefore particularly susceptible to the problems associated with large scale.

The larger the organisation the more complex are problems of coordination and communication. The best decisions are stultified if they are not properly communicated, but a flow of orders in an attempt to ensure adequate communication tends to lead to red tape, a stifling of initiative and an attitude of unwillingness to take responsibility. Matters are dealt with according to rule and precedent instead of on their merits. (Civil servants are perhaps particularly susceptible to these deficiencies because they are accountable to Parliament through their ministers for their actions. They therefore wish to cover themselves as far as possible by being able to quote precedent and rule in their defence.)

The larger the organisation the greater the danger of remoteness of control, of delay, bottle-necks and an executive out of touch with the urgencies of the real situation. Also, the larger the organisation the more difficult it is to keep the different parts in step.

Let us take one particular, one might indeed say notorious, example of a failure of coordination in the public sector. It arises with the so-called 'poverty trap'.

There have grown up over the years a series of means-tested benefits. On the last count there were over forty such benefits, some introduced by local authorities, others by government departments. Each has been designed to try to improve the lot of the poor in some particular respect or to shield them from some new charge about to be imposed. One of the most important of these provisions is the Family Income Supplement (FIS), which is restricted to married men with children who are in work but on low pay. They receive half the difference between a set scale, based on size of family, and their own wage. There are also rent rebates (for council house tenants) and rent allowances (for tenants in private property), rate rebates and means-tested benefits for school meals and milk.

The essence of the means-tested benefit is that it goes to those whose income is low and, necessarily, if and as income increases, so the benefit is withdrawn. Thus, a worker receiving FIS would suffer a 50p reduction in FIS if his income rose by £1 per week.

These various means-tested benefits have not been coordinated with each other nor with the income tax system. The income threshold at which some of the benefits become applicable is above the threshold at which income tax has to be paid. Also, those in work pay National Insurance contributions. If, then, a poor person's income rises by £1 a week he will lose part of this increase in the withdrawal of benefits and at the same time will be paying between 30 and 40 per cent of it in additional income tax. The net result is that he may actually find himself worse off from having had an increase in gross income. Even where he is still left with something, the combined effect of loss of benefit and tax may be, say, a 90 per cent effective combined rate. This is sometimes referred to as the 'poverty surtax'.

The implications are clear. Poverty is not being adequately relieved; there is a strong disincentive to work; wages policies designed to help the lowest paid may be frustrated; but, above all, the poor are put in the position where it is almost impossible for them to drag themselves out of poverty by their own efforts. One must beware of overstating the effects of the poverty trap because not all the reductions in income associated with the withdrawal of benefits apply immediately. There are different time periods to establish eligibility and it may be some time before all the reductions in income have taken place. In that sense the psychological impact is not as great as we have suggested. But the problem is a very real one; and it arises basically because of a lack of coordination between the large departments of national government, in particular, a lack of coordination between the social security system and the tax system, but also, to a less extent, between central and local government. The problem has been recognised for over a decade but, once having been allowed to come into existence, the stage has been reached at which eliminating the overlap between social security and income tax would require either an unacceptable fall in the threshold for benefits or a very expensive increase in tax thresholds.

Let us turn now to another aspect, the motivations of bureaucrats and the effect of these motivations. The analysis of the economics of bureaucracy owes much to Niskanen (1973). He argues that just as entrepreneurs seek to maximise their profits and consumers to maximise their utility, so bureaucrats seek to maximise their budgets. He defines a bureau as a unit of an organisation which receives part of its income by grant rather than by sale of its services. On this definition bureaux may occur in private industry but they are predominantly to be found in the public sector. A bureaucrat on his definition is particularly someone who is in charge of a bureau. He then asks the question: 'What are the motivations of bureaucrats?' He finds the answer in a variety of advantages such as increased salary, perquisites of office, power, influence, prestige, patronage, all of which are likely to be promoted by an increase in the size of the bureaucrat's budget. Even if the bureaucrat goes in for an easy life, in the short run at any rate, life is likely to be made easier for him if he has more funds at his disposal. Thus budget maximisation can be taken as a valid proxy for the motivations of bureaucrats.

To say that bureaucrats seek to maximise their budget does not necessarily imply a cynical view of their activities. Many have a tremendous pride in the output of their bureau. They are dedicated to their work and identify an increase in their 'service', be it defence, education or health, with the public interest. Even if a bureaucrat considers that some of the funds he uses might in principle be better spent elsewhere, he may have no confidence that this will happen if he foregoes them.

There are perhaps three main conclusions that may be drawn from this analysis. The first is that the bureaucracy provides a built-in upward pressure on spending in the public sector. Not only that, in some measure, the principles on which the British Civil Service has been assumed to operate are undermined. The assumption has been that civil servants are there to serve and carry out the policy of ministers. Where the policy is one of expanding the public service this may be true; but where the ministers seek to cut back public expenditure, then they are more likely to run into the opposition of senior bureaucrats acting individually, and junior bureaucrats acting through their trade unions, to defend their jobs and conditions of work.

Secondly, bureaux are less likely to be concerned with economising than profit-making organisations. If a bureaucrat reduces costs by 5 per cent he may have that much more to spend on other items. But if the head of a profit-making organisation, which previously had a 5 per cent difference between receipts and costs, saves 5 per cent on costs, profits rise by 100 per cent. Moreover, the bureaucrat does not face the pressures of a competitive situation. A profit-making organisation that fails to economise may simply not survive. The same fate rarely applies to a bureau.

Thirdly, this analysis raises the whole question of motivations in society. We referred, at the end of Chapter 3, to the view that the market system was abhorrent to some people because of its basis of self-interest. It is sometimes implicitly assumed that the State has available to it a fund of wisdom not found in the rest of the community, and that once responsibility for a measure is passed over to the State, the measure will be pursued with disinterested public spiritedness. In fact civil servants who administer public policy are men and women like the rest of us. They have no superior fund of wisdom and they are not disinterested. At the moment of writing civil servants are locked in an industrial dispute with the government over wages and methods of pay settlement; have been engaging in a series of selective strikes which have caused considerable inconvenience and loss to some groups in the community; and are threatening an all-out strike in which they would refuse to pay unemployment benefits. Of course, duty and public-spiritedness are characteristics that are to be found in the Civil Service; but it may be questioned whether they are more prevalent there than elsewhere in society.

8
THE SCOPE AND LIMITS OF GOVERNMENT POLICY

The range of policy measures

One fact should stand out clearly from the foregoing chapters: the economic structure is closely allied to the political structure; to an important extent economics and politics go hand in hand; it is impossible to consider the economic structure meaningfully without reference to the political structure.

In the mixed economy of the United Kingdom it is also clear that the range of issues with economic policy implications is wide indeed. We have looked at some, and mentioned others only in passing. They include industrial policy: the outlawing of unfair competition, the control of private monopolies, the right policies for the nationalised industries and how to deal with the trade unions. They cover the use of resources to provide pure public goods, like defence and law and order, and policy on quasi-public goods, like health and education, with their externality effects. Policy on poverty needs to be improved to try to abolish the poverty trap. Economic policy also includes taxation, and it affects and is affected by the relationship between central and local government authorities. Although barely mentioned in this text, government economic policy likewise comprehends action on pollution and decisions about the use of exhaustible materials. One could go on. There are, however, two areas of policy which require particular attention. Both relate to important deficiencies or failures in the market system: the generation of inequities in the distribution of income and wealth; and the problem of unemployment, linked to that of inflation. The question is, how far can the defects be rectified by State action without at the same time undermining the merits of the market system, possibly leading to its replacement by something worse? It is to these two issues that we now turn.

The distribution of income and wealth

As we saw in Chapter 3, the market system generates inequality in the distribution of income and wealth. Some people may have incomes which are very much larger than others not because they work harder or longer, but because they happen to possess abilities which, to their good fortune, are held in high demand by society and hence can command a high market price. Further, under the influence of changing technologies and changing tastes, market demands for resources change and the price of labour and of capital in particular uses therefore changes. These price changes are a vital part of the efficient working of the price system to re-allocate resources in accordance with society's needs, but they bring fortuitous increases in income to some and equally fortuitous income reductions to others. Differences in earned income provide unequal opportunities for saving and different amounts of savings in turn generate differences in investment income. Moreover, where wealth can be transmitted between generations, inequality may not only be perpetuated but accentuated. To many, gross inequalities of income and wealth are highly objectionable. That some who may have done nothing to merit it should have much, while others who may have done nothing to deserve it should have little, is offensive and, indeed, unacceptable. How serious is the problem in the United Kingdom and what can government do about it?

The most comprehensive analysis of income and wealth distribution in the United Kingdom is to be found in the series of reports published by the Royal Commission on the Distribution of Income and Wealth (the Diamond Commission). Report No. 7 (Cmnd 7595, July 1979), the last of the Commission's reports to review the general distribution of income and wealth, takes the picture to 1977.

We can examine income distribution by source and among persons. On distribution by source the report shows that over the ten years from 1967 to 1976 the share of income from employment and self-employment together (including occupational pensions) varied between 78 and 80 per cent of total personal income (with income from self-employment averaging about 9 per cent). In 1977 income from employment was its lowest for eleven years at 77.9 per cent. The fall was balanced largely by an increase in transfer income which reached a peak of 12.1 per cent in 1977. Rents dividends and interest accounted for the residue (some 10 per cent).

As so often, the statistics are imperfect and the imperfections are more significant when we look at income distribution among persons. Some of the main limitations of the data are that the 'income unit' is not the individual but the income unit for tax purposes (married couples generally counting as one unit); coverage of employee fringe benefits is incomplete; there is some under-recording of income; and capital gains do not count as income, though many economists think they should be so treated. However, the figures we use are compiled on a consistent basis and, as the Diamond Commission puts it, the distributions 'while not as comprehensive as we would wish, provide worthwhile indicators of inequality and of trends in inequality' (p. 7).

The usual way of showing the distribution of income among persons is by indicating the proportion of income received by different percentiles of the population. The situation is summarised both pre- and post-income tax in Table 9, which compares 1976–7 with 1949.

The table suggests a substantial decline in the pre-tax share of

TABLE 9 *Distribution of income by personal shares, 1949 and 1976–7*

	Pre-income tax		Post-income tax	
	1949 (%)	1976–7 (%)	1949 (%)	1976–7 (%)
Income share of top				
1 per cent	11.2 ⎫	5.4 ⎫	6.4 ⎫	3.5 ⎫
Income share of next	⎬ 33.2	⎬ 25.8	⎬ 27.1	⎬ 22.4
9 per cent	22.0 ⎭	20.4 ⎭	20.7 ⎭	18.9 ⎭
Income share of next				
40 per cent	43.1	49.7	46.4	50.0
Income share of bottom				
50 per cent	23.7	24.5	26.5	27.6
	100.0	100.0	100.0	100.0

Source: Cmnd 7595 (1979), pp. 16 and 24.

both the top 1 per cent and the top 10 per cent of income receivers during the period between 1949 and 1976-7;[1] but the gainers were primarily in the top half of the distribution. There was very little increase in the share of the bottom 50 per cent of income receivers. Post-tax the share of the top 1 per cent is further heavily reduced in both years and this time the bottom 50 per cent are substantial gainers. Comparisons with years before 1949 are difficult, but data for 1938 suggests a much more unequal distribution than in 1949.

For the full effect of government policy on real income distribution we need to take account of taxes on goods and services and of the distribution of the benefits of government expenditure other than transfer payments. These are difficult to allow for with any precision, but the available evidence suggests that the effect is a substantial reduction in inequality.

Some significant differences exist in the distribution of the different sources of income by personal shares. The most outstanding is investment income, which is much more unevenly distributed than income from employment. In 1976-7 the top 10 per cent of income receivers pre-tax received over a half of all investment income, and the top 1 per cent nearly one-quarter.

How far can a government reduce income inequalities by means of income tax without running into serious economic problems? One fear is that high rates of income tax will have disincentive effects - on willingness to work, save and take risks; in economists' terms, that it will reduce the supply of the factors of production. However, income tax does not necessarily have this effect. Take the case of labour. An increase in income tax will have two opposing effects. On the one hand (assuming the government does not spend the extra revenue in ways which precisely compensate him) a taxpayer will be worse off in terms of the particular goods and services he buys. He can now afford less of those things he enjoys, including leisure; he will tend to have less leisure, i.e. to do *more* work, an incentive effect. On the other hand, the tax has decreased his net earnings from an hour's work. If he does an hour's less work, he sacrifices less income (and the things income can buy) than before. On this count he is likely to substitute leisure for work, a disincentive effect. Thus, from an increase in income tax there is both an income effect and a substitution effect, which work in opposite directions.

It may seem that this analysis brings us no nearer to an answer, but there is an important difference between these two effects. The income effect works through the change in total income while the substitution effect works through the change in the terms on which income can be acquired at the margin. The particular form an income tax may take for any given revenue yield affects the relative strength of the two effects. Thus an income tax which took the form of a lump sum tax on all income receivers irrespective of the size of their income (in effect a 'poll' tax), would have an income effect but no substitution effect: it would be an incentive. It would also be very regressive. Conversely, an income tax with very high marginal rates, which was very progressive, would be more likely than a proportional tax to have a disincentive effect.

Beyond this, we must look at the studies of how workers behave, or say they behave, in response to income tax. There have been a number of such studies.[2] From their findings it would appear that at the level of basic rate there is, if anything, a net incentive effect of income tax; but at the highest tax rates there is some evidence of a net disincentive effect. There are also some indications that high tax rates discourage married women from taking paid work other than on a part-time basis which keeps their income below the threshold of the married women's earned income relief.

Similar 'income' and 'substitution' effects apply in relation to income from saving and investment.

The arguments on incentive do not, however, dispose of the economic problems associated with high taxation. Although the total effect on the supply of labour and of savings for investment may not be large, there are distorting effects in the form of tax avoidance and evasion. Tax avoidance is the adoption of legal methods to reduce one's tax bill and evasion the adoption of illegal methods. Thus salary earners may protect their real income against tax by having more of it in the form of payments in kind which are either untaxed or not taxed to the full extent of the benefit. Moonlighting wage-earners may join the 'black economy'.

Perhaps the most serious economic effects of high marginal tax rates in the United Kingdom in recent years has been on the pattern of investment. We noted earlier (Ch. 6) the decline of private investment in ordinary shares. Individuals have taken advantage of tax reliefs to put their money into housing, insurance policies and pension schemes instead of ordinary shares. Again, if tax rates became too high then investors may seek investments yielding capital gain rather than income. In the last resort this might mean buying antiques, paintings or stamp collections which can be expected to appreciate, yield no income taxable to income tax, and, if held until death, avoid all capital gains tax.

Before the Conservatives came to office in 1979 the maximum marginal rate of income tax on earned income was 83 per cent; on substantial investment incomes there was a further tax (the investment income surcharge) of 15 per cent, making a maximum marginal rate of income tax of 98 per cent. Thus if a rich investor obtained an extra £100 of investment income he was allowed to keep only £2 of it. This was an absurd level of income tax if only because it was self-defeating: almost everyone threatened by it avoided it. The present maximum level of 60 per cent for earned income and 75 per cent for investment income is probably as high as it is sensible to go.

There are, therefore, certain limits beyond which income tax cannot be pushed without adverse effects; but considerable redistribution is possible before these contraints become serious. In fact, the distribution of income in the United Kingdom is now among the least unequal of those advanced countries with which comparison can be made (Cmnd 6999, 1977).

A more fruitful field for policy measures to reduce inequality is in relation to wealth. Wealth is more unequally distributed than income and measures which reduced inequality of wealth would also strike at the source of the massive inequality in investment income that we have noted.

What is personal wealth? It consists of assets of all kinds: money in the bank and under the bed; personal possessions from clothes to cars; houses and land; business assets like machines or office buildings owned by entrepreneurs or partners; business assets in companies in the form of shares held by shareholders; government bonds and other forms of government debt in private hands.

There remain some difficult borderline cases. One such is the value of pension rights. If a man is entitled, on retirement, to a pension for life, ought not the value of this right to count as part of his wealth? Because of his pension right he needs to accumulate less wealth than otherwise he would have to. There is a strong case for saying that it should so count. On the other hand it is not marketable wealth. It cannot be sold and (subject to some possible rights of transfer to a widow) it is not transferable. It is wealth of a rather different order from marketable wealth.

In examining the distribution of wealth we must again recognise limitations to statistics; but there is no reason to believe that the deficiencies seriously distort the picture (Atkinson and Harrison 1978). The figures for 1976, the latest year covered in the Diamond Commission's statistics on the wealth owned by those aged eighteen or over, are set out in Table 10. Column 2 shows the distribution of marketable wealth only, while columns 3 and 4 show estimates including the value of occupational pensions and State pensions.

It can be seen that the top 1 per cent of wealth holders owned about one-quarter of marketable wealth; this figure fell to 21 per cent when occupational pensions were included and to 14 per cent when State pensions were also added. Conversely, the bottom 80 per cent of wealth holders owned only 22 per cent of marketable wealth, but 45 per cent of wealth including all pension rights.

Table 11 shows trends in the distribution of marketable wealth for selected years of the period 1966-76. There are some differences in comparability but they are not sufficiently serious to distort interpretation of the trend. The picture to emerge is of a significant fall over the period in the wealth held by the top 1 per cent of

TABLE 10 *Distribution of personal wealth, United Kingdom, 1976*

Quantile group	All marketable wealth (pension rights excluded) (%)	Wealth including occupational pension rights (%)	Wealth including occupational and State pension rights (%)
Top 1 (%)	24.9	21.1	14.1
2-5	21.3	20.1	15.0
6-10	14.4	14.0	11.2
11-20	17.0	17.4	15.0
21-100	22.4	27.4	44.7
Cumulative basis			
Top 1 (%)	24.9	21.1	14.1
5	46.2	41.2	29.1
10	60.6	55.2	40.3
20	77.6	72.6	55.3

Source: Cmnd 7595 (1979), Tables 4.3, 4.13 and 4.15.

TABLE 11 *Trends in the distribution of marketable personal wealth; selected years 1966–76*

Quantile group	1966 (GB) (%)	1970 (GB) (%)	1974 (UK) (%)	1975 (UK) (%)	1976 (UK) (%)
Top 1 (%)	31.1	29.4	22.5	23.5	24.9
2–5	24.3	22.5	20.6	20.3	21.3
6–10	13.1	14.2	14.4	14.2	14.4
11–20	—	—	18.8	18.2	17.0
21–100	—	—	23.7	23.8	22.4
(11–100)	(31.5)	(33.9)	(42.5)	(42.0)	(39.4)

— = not available.
Source: Cmnd 7595 (1979), Table 4.4.

wealth holders; a more modest fall in the wealth holdings of the top 2–5 per cent; and a rise in the holdings of the other quantile groups including the top 6–10 per cent.

Although comparisons with earlier years are unreliable in detail because of differences in coverage it is clear that the pattern since 1966 continues an earlier trend. Estimates for England and Wales for 1923 show as much as 61 per cent of marketable wealth held by the top 1 per cent; 21 per cent held by the top 2–5 per cent, and 7 per cent by the top 6–10 per cent. It is clear that over this century there has been a substantial reduction in wealth inequality in Britain; but much of the wealth lost to the top 1 per cent of wealth holders remained within the top 10 per cent.

There are strong reasons why any policy to reduce inequality in the distribution of wealth by means of taxation should do so by taxation of property passing at death (allied to a gift tax to prevent avoidance). It is administratively easier than an annual wealth tax as only a proportion of total wealth becomes liable to tax in any one year, and then at a time when an inventory and valuation of the property are required anyway for purposes of carrying out the will or implementing the law of intestacy. Any adverse economic effects are likely to be minimised. As the burden falls primarily on the heirs, a tax at death does not penalise those who have increased wealth by saving and enterprise. Moreover, the detailed researches of Professor Harbury (Harbury and Hitchens 1979), comparing the estates of parents and children at death, have made it clear beyond all reasonable doubt that inheritance is the biggest single source of inequality in the distribution of wealth. Finally, provided that the position of widows is safeguarded, inheritance is a particularly suitable subject for taxation because the beneficiaries are receiving wealth which is usually unrelated to their own effort or merit. In this respect it should do least offence to the private enterprise ethic – that rewards from the economic system should be proportional to contribution.

The most effective form of death duty in reducing inequality is one where tax is imposed on what the beneficiary gets (usually called an inheritance tax) as against a tax imposed on what the deceased leaves (an estate tax). In the UK the present capital transfer tax, like the previous estate duty, is imposed on the total value of the estate left at death. Thus the amount of tax is the same whether a millionaire leaves all his wealth to one person or to be divided equally among 100 people. But under a progressive inheritance tax the sole heir of a millionaire would pay heavier tax than under estate duty while the 100 beneficiaries together would pay considerably less. An inheritance tax can be expected to reduce inequality in the distribution of wealth in two ways. It encourages the wealthy to spread their bequests because, by so doing, they reduce the total amount of tax to be paid. More significantly, however, it is large receipts, not large estates as such, which *perpetuate* inequality and an inheritance tax therefore strikes at the heart of the matter.

Unemployment and inflation

Historically, economies with predominantly market sectors have been marked by cyclical fluctuations in output, prices and employment of about ten to twelve years' duration. Between the two world wars the cyclical fluctuations continued but with an overall higher level of unemployment. Between the boom at the end of the First World War and the rearmament just before the Second World War the level of unemployment in the UK never fell below 10 per cent of the insured population and in the Great Depression of 1929–33 the level of unemployment at its peak was over 20 per cent.

The work of Keynes and others in the late 1930s and early 1940s led to the view that unemployment could be cured by State action, and in a White Paper in 1944, supported by all the political parties, the coalition government accepted the obligation to seek to maintain a 'high and stable' level of employment. This responsibility was to be carried out mainly by budgetary policy (changes in public spending and taxation) and to a lesser extent by monetary policy (changes in the quantity of money and the rate of interest).

Optimism that the State could succeed in such a policy arose from the nature of the Keynesian analysis of unemployment. While unemployment might arise in individual industries from a variety of causes, such as changes in technology, *general* unemployment was a product of deficiency in the total demand for goods and services. This deficiency may arise because the rate of interest fails to bring into equilibrium the amounts savers wish to save with the amount that investors wish to invest in plant and equipment. If people wished to save more than others wished to invest, then the result would be a reduction in consumption demand without a corresponding increase in demand for investment goods. Total demand would fall and workers would become unemployed.

Where this situation threatened, government could intervene by increasing effective demand either by expanding its own expenditure without raising additional taxes to match or by lowering taxes, and thus increasing private demand for goods and services, without cutting its own expenditure. If aggregate demand became so high as to generate unacceptable price rises, then the government could respond by reverse measures to those promoting employment; it could cut its own expenditure without cutting taxes or reduce private demand by increased taxes without increasing its own spending.

This budgetary policy was implemented in the post-war period and, between the end of the Second World War and the late 1960s, measured by levels of unemployment, it worked well: only for relatively short periods did unemployment rise above 2 per cent of the insured population. However, the less favourable side of the coin was that the period saw a continuous rise in prices averaging about 4 per cent per annum. Not only did the price rise have unfortunate social consequences, such as drastically reducing the value of savings in the bank or in government securities, the inflation played a significant part in recurrent balance of payment crises and was the root cause of the two devaluations of 1949 and 1967.

However, much worse was to come. The decade of the 1960s

ended with a new phenomenon which did not fit into the postwar orthodoxy: both rising prices and rising unemployment at the same time. And the 1970s saw levels of unemployment and rates of price increase far higher than anything in the previous period. In round terms prices doubled between 1966 and 1975 and the rate of inflation accelerated in the 1970s, reaching a peak between January 1974 and February 1976, when the index of retail prices increased by 50 per cent. Unemployment rose to over a million, fell back somewhat, and then rose again under the Labour Government of 1974 and continued to rise under the Conservative Government of 1979 to reach a figure of over 10 per cent of the working population by 1981. At the same time year-on-year price increases were still around double figures.

There are two broad approaches to remedying this situation. The first sees the main cause of inflation as pressure for wage increases exerted by powerful trade unions who, because governments are committed to a full employment policy, are not deterred from pressing claims by the fear of pricing their members out of jobs. The three desirable objectives of full employment, a stable price level and free collective bargaining are incompatible. Thus a solution must be found by some restrictions on trade union action with perhaps the acceptance of a modest rate of inflation and possibly a slightly higher level of unemployment than in the first decade after the Second World War. This policy requires wage restraint. Successive governments have sought to employ voluntary or compulsory incomes policies but they have achieved very limited success. In the short-term they may have held back wages, but when the policy has been relaxed the dams have broken, and bigger wage increases have followed to make up. Moreover, such policies have created inequities which have led to social and industrial tension. For example, the imposition of arbitrary wage freezes means that some workers will have just received a big increase while others, who were about to get one, are denied any increase at all. Moreover, incomes policy undermines the efficient working of the price mechanism. If restrictions on wages and, indeed, on other incomes are imposed across the board, movements in relative wages cannot act to bring about the redistribution of labour which a dynamic and expanding economy requires.

The failure of these incomes policies has led to a search for an alternative approach; theoretical backing has come from economists known as monetarists. They argue that trade unions do not cause inflation; rather, inflation is caused by an excessive increase in the quantity of money. If the growth in the money supply is kept broadly in line with the growth in the real national output, then inflation will be checked. Moreover, until inflation is checked, no policy for stable economic growth or a permanently higher level of employment can be successful. Monetarists put much stress on expectations and curing inflation requires the elimination of inflationary expectations. It is this approach which the Conservative Government of 1981 is following. The government is seeking to control the money supply more tightly than hitherto and to cut public spending and borrowing in an attempt to squeeze inflation from the system.

It should be acknowledged that the problems of the past decade or so have not been solely domestic ones. Since 1973 the world economy has been submitted to intense strains, mainly because of the huge increase in oil prices; and the late 1970s and early 1980s have seen high rates of inflation and unemployment in most mixed economies. However, the UK experience has been among the worst.

Will the Conservative Government strategy succeed? Most economists agree that by a sufficiently tough monetary policy it is possible to get rid of inflation, but the strategy has flaws. First, money itself is not easy to define or control; second, the elimination of inflation by monetary policy may require a wholly unacceptable level of unemployment in the short run and perhaps do permanent damage to the industrial structure; third, once inflation is removed and a stimulus again given to production, it is by no means certain that the wage/price spiral will not start up again.

In the latter case we would be forced back on an incomes policy. It may be possible to design one which is strong enough to control inflation and does not undermine the market system and the outline of one such policy has been suggested by Professor James Meade (1971). He has proposed that the government might set a norm for the annual percentage rise in earnings, but that any group of employers and employees would be free to reach agreement on a wage bargain whether or not it was above the norm. If no agreement was reached the matter would be referred to a tribunal whose responsibility would be to determine simply whether or not the wage demand exceeded the norm. If the tribunal ruled that it was above the norm, workers who tried to enforce their claim by industrial action, such as a strike, would be subject to sanctions, e.g. a loss of accumulated rights to redundancy pay or a tax on strike money paid out by their union. Where the claim was ruled *not* to be above the norm, no curbs would be placed on the bargaining powers of the union concerned.

Were this procedure to be introduced at a time when the rate of inflation was very high, say 20 per cent per annum, the norm might initially be set at 15 per cent and subsequently reduced to 10 per cent, and so on.

Such a policy would restrain wages in general, at the same time enabling the price system to fulfil its efficiency functions. If, because of a change in economic conditions such as a shift in demand or a technological innovation, there were a genuine shortage of workers in a particular occupation, employers and employees in that industry would agree an above-norm increase.

In Chapter 1 we spoke about some of the inherent complications of the social sciences arising from their concern with human behaviour. On this topic of unemployment and inflation our understanding of behaviour is incomplete. Government policy suffers from lack of knowledge and economists are divided about the fundamental causes of inflation and in their interpretations of how the economy will react to certain changes. Unfortunately, in these circumstances, good intentions are not enough for a government.

The social balance

Let us conclude by considering the question of where the social balance should lie. What should be the respective sizes of public and private sectors? The vast majority of people in the United Kingdom wish to support the mixed economy; they desire neither to return to an economy which is overwhelmingly market or to advance to an economy which is overwhelmingly State; the failures of the market and the failures of the State are too marked for that. But what is the optimum mix?

To that question there is no precise answer. It rests on value judgements not only about the extent of State intervention but about its form. On the one side are those who are drawn to a more individualist philosophy – who lean towards personal and family decision-making and see moral value in personal development and personal choice. On the other hand are those drawn to a more collectivist philosophy who lean to increased social provision and who are attracted by the idea of common citizenship and building up a sense of community.

Some policies will please them both. In areas where market provision is not in dispute, then both will applaud measures to eliminate unfair competition and to increase the knowledge available to consumers. Such provisions make private markets work better. But on many issues the individualists and collectivists will divide.

Thus the individualist will be more tolerant of market failure than the collectivist. Where he considers the deficiencies of the market need correcting he will seek methods of doing so which leave more choice to the individual. He looks towards a freer and less restricted milieu for industry, and is unhappy about monolithic nationalised industries. He leans to efficiency and liberty rather than equality. He will believe in promoting equality of opportunity, but may not wish to go as far as the collectivist in reducing the inequalities of income and wealth which emerge from the market process. In terms of measures he leans to finance rather than provision; he would wish to see inequalities in income and wealth reduced by taxation and transfers rather than a series of equalising services. He would wish to retain a private sector for education and health using vouchers to ensure minimum access to all.

The collectivist on his side is more tolerant of State failure than of market failure. He would see tighter restrictions on industry and is especially suspicious of multi-national companies. He dislikes vouchers as socially devisive and tends to favour universal State education and a universal National Health Service as providing a more congenial social milieu and a less class-ridden society. He is not as unhappy as the individualist about large nationalised industries, although he would seek ways of making them more efficient and more democratic.

This picture, of the difference between individualist and collectivist, may be somewhat overdrawn. Certainly not all people fall clearly into one camp or the other. Nonetheless these are two vital strands of thought affecting which way the society goes. It is not for the economist, as such, to make a judgement between the two philosophies. His task is to try and draw out the implications of a movement in one direction or another; to show what is the cost in terms of resources, and to indicate whether there may not be a better way of securing any particular objective. However, which way the community goes will profoundly affect the economic structure, and the political structure, and indeed the kind of society in which we live.

NOTES

Chapter 1 The science of economics

1. Government stock or bonds generally carry a fixed rate of interest calculated in relation to the nominal value. The maturity date (the date at which the government promises to repay the loan) varies with different bonds from a year or so from issue at one extreme to the irredeemable (where the government never undertakes to repay) at the other. It may be asked: 'Why on earth should anyone be willing to make a loan which is never going to be repaid?' The answer is that the lender, in purchasing the bond, obtains the right to a specified annual money income (the interest) in perpetuity. Moreover, this right is a saleable asset, so that should the bond holder need cash he can readily sell his bond on the market, though with no guarantee as to price. In fact changes in the price of an irredeemable bond exactly reciprocate changes in the rate of interest. Suppose a bond was issued 'at par', that is at £100, with the market value equal to the nominal value, carrying an interest rate of 5 per cent. This means that the government undertakes to pay £5 per annum for every £100 bond. At a later date suppose the government wants to borrow a lot more money, so much that in order to find sufficient lenders it has to offer 10 per cent. At this point the price that can be obtained by a bond holder of the 5 per cent issue will be only £50. People will not pay more than £50 for the right to receive £5 per annum if they can obtain £10 per annum from a newly issued bond which costs £100. The rate of interest has doubled, the price of the bond has halved. Securities which have a definite redemption date will also vary in price inversely with changes in interest rates but not to the same extent as an irredeemable bond. The nearer the bond to the maturity date the less the change in its price following a given change in interest rates.
2. For examples of pitfalls in the use of statistics, see C. T. Sandford, *Social Economics*, 1977, Heinemann Educational Books, Ch. 2.

Chapter 5 The mixed economy: the United Kingdom

1. There have been fashions in the official statistics used to express public spending as a proportion of national output. The figures for the later years in Table 2 are derived from the annual *National Income and Expenditure* Blue Books and match up as far as maybe with the earlier figures prepared by Veverka (1963). Recently, in the annual Public Expenditure White Papers, the statisticians have been using a narrower definition of public spending (the biggest difference being the exclusion of all interest on public debt against which there is a corresponding income, e.g. rent from council houses). Following international practice the

government has also switched to a market price measure of GNP. On this basis the figures of public expenditure for the financial years since 1971 are as follows:

Year	% of GNP	Year	% of GNP
1971-2	38.0	1976-7	44.5
1972-3	39.0	1977-8	40.5
1973-4	41.0	1978-9	41.5
1974-5	46.5	1979-80	41.5
1975-6	46.5	1980-1 (estimates)	44.5

Figures for earlier years have not been published on this basis. It can be seen that, in absolute terms, this measure results in considerably smaller figures than those in Table 2.

Chapter 6 The private sector in the United Kingdom

1. The term 'entrepreneur' was borrowed from the French because of the lack of an English equivalent. 'Sole trader' won't do because it implies a one-man firm and suggests commercial activities only. 'Undertaker' has been used by earlier economists as a translation of the French word, but its somewhat morbid associations make it unacceptable.
2. One safeguard for shareholders against inefficient directors is the possibility of a takeover. If the firm's assets are not being well used some person or organisation will see the prospect of profit from taking over the firm and changing the board of directors and policy. The takeover bidder will seek to acquire a sufficiently large number of shares to be able to vote out the directors. The shareholders will benefit both from the rise in the price of their shares as a result of the bid and from the new policy, if it is successful.
3. This section on retailing draws heavily on the excellent and comprehensive article by J. A. N. Bamfield 1980.
4. Clarke (1979) records an estimate that well over 50,000 million payments are made in Britain every year; that nearly 95 per cent of these payments are made in cash and virtually the whole of this money is distributed by the clearing banks; and of the rest (money transmitted in the form of cheques and credit cards) the clearing banks handle some 80 per cent (p. 10).
5. A reminder of the time when bank notes were backed by gold is to be found in the words still printed on them, 'I promise to pay the bearer on demand the sum of one pound', signed by the Chief Cashier of the Bank of England. If you took your bank note to a clearing bank or to the Bank of England and demanded a pound in exchange, all you could get would be another note or one pound's worth of the usual coins.

Chapter 8 The scope and limits of government policy

1. There is some evidence that top salary earners may have partly compensated for their decline in money earnings by increased non-cash benefits. (See Diamond Commission, Cmnd 7595, p. 51.)
2. For a fuller treatment of this topic including an account of a number of the empirical studies see C. T. Sandford, *The Economics of Public Finance*, Pergamon, 2nd edn (1978), Ch. 6.

REFERENCES AND FURTHER READING

A. B. Atkinson and A. J. Harrison, *Distribution of Personal Wealth in Britain*, Cambridge University Press, 1978.

R. Bacon and W. Eltis, *Britain's Economic Problem: Too Few Producers*, Macmillan, 1976.

J. A. N. Bamfield, 'The changing face of British retailing', *National Westminster Bank Quarterly Review*, May 1980.

W. M. Clarke, *Inside the City*, George Allen & Unwin, 1979.

P. V. Elst, *Capitalist Technology for Soviet Survival*, Institute of Economic Affairs, Research Monograph 35, 1981.

J. K. Galbraith, *The Affluent Society*, Pelican, 1962.

C. D. Harbury and D. M. W. N. Hitchens, *Inheritance and Wealth Inequality in Britain*, George Allen & Unwin, 1979.

J. R. Hicks, *The Social Framework*, Clarendon Press, 4th edn, 1971.

G. Jones, 'Some post-Layfield reflections about non-domestic rating in commercial and industrial properties' in R. Jackman (ed), *The Impact of Rates on Industry and Commerce*, CES Policy Series No 5, Centre for Environmental Studies, London, 1978.

N. Kaldor, *Causes of the Slow Rate of Growth of the United Kingdom*, Cambridge University Press, 1966.

Layfield Report, *Report of the Committee of Inquiry into Local Government Finance*, Cmnd 6453, HMSO, 1976.

C. E. Lindblom, *Politics and Markets*, Basic Books, 1978.

J. E. Meade, *Wages and Prices in a Mixed Economy*, Occasional Paper 35, Institute for Economic Affairs, London, 1971.

W. A. Niskanen, *Bureaucracy, Servant or Master?*, Institute for Economic Affairs, London, 1973.

A. T. Peacock and J. Wiseman, *The Growth of Public Expenditure in the United Kingdom*, George Allen & Unwin, 1967.

E. H. Phelps Brown, 'The underdevelopment of economics', *The Economic Journal*, 325 **82**, March 1972, p. 10.

A. R. Prest and D. Coppock (eds), *The UK Economy: A Manual of Applied Economics*, Weidenfeld & Nicolson, 8th edn, 1980.

R. Pryke, *The Nationalised Industries: Policies and Performance since 1968*, Martin Robertson, 1981.

Royal Commission on the Distribution of Income and Wealth, *Report No. 5*, Cmnd 6999, HMSO, 1977.

Royal Commission on the Distribution of Income and Wealth, *Report No. 7*, Cmnd 7595, HMSO, 1979.

C. T. Sandford, *Social Economics*, Heinemann, 1977.
C. T. Sandford, M. Godwin, P. J. W. Hardwick and I. Butterworth, *Costs and Benefits of VAT*, Heinemann, 1981.
Survey 'How you rate your jobs', *Money Which?*, September 1977.
A. Sutherland, 'Capital transfer tax and farming', *Fiscal Studies*, March 1980.
J. Veverka, 'The growth of government expenditure in the United Kingdom since 1790', *Scottish Journal of Political Economy*, vol. X, No. 1, 1963.

INDEX

Acton, Lord, 43-4
advertising, 34, 40
agriculture, 18, 56, 57, 61-3, 76
Atkinson, A. B., 96, 107

Bacon, R., 65, 107
balance of payments, 21, 99
Bamfield, J. A. N., 65-6, 106, 107
bank (banking), 15, 56, 67-70, 71, 85, 86
   Barclays, 68, 70
   clearing, 68-9, 85, 106
   Lloyds, 69
   Midland, 69, 70
   National Westminster, 68
Bank of England, 68, 69, 75, 106
   powers of, 85-6
barter, 19, 39, 45
benefit, 23, 31, 41, 49, 53, 79
   child, 78
   indivisible, 32
   means-tested, 86-7
   private, 32, 33, 40
   social, 32, 33, 40
   social security, 6, 88
   supplementary, 78
   unemployment, 72, 79, 89
British Airways, 83
British Leyland, 83-4
British Rail, 83-4
British Steel Corporation (BSC), 83-5
bureaucracy, 86
   economics of, 75, 88-9

capital, 6-8, 11, 13, 14, 26, 29, 34, 37, 57, 58, 59, 61, 69, 78, 85, 91
capital gain, 92, 95
capitalism, 44

Census of Production, 63
charges, 31
   for local government services, 79, 81, 82-3
China, 39
City, The, 67-8, 70
Clarke, William, 68, 106, 107
Clarks, C. and J., 59
collectivist philosophy, 103-4
competition, 40, 69, 75, 83, 85, 89, 90, 103
   imperfect, 33
   perfect, 33
Conservative
   Government, 11, 54, 73, 80, 101
   Party, 81, 82
consumer, 27, 28, 29, 36, 37, 42, 53, 103
consumption, 13, 14, 30, 32, 46
   capital, 24
cooperatives, 61, 66
Coppock, D., 107
cost, 12, 23, 27, 28, 29, 32, 41, 42, 46, 49, 60, 67, 89, 104
   average, 12, 63, 64
   factor, 21, 25, 49
   opportunity, 10-11
   private, 32, 33, 40
   social, 32, 33, 40
creditor nation, 22, 25
Cuba, 39

death duties (taxes), 62, 97-8
   capital transfer tax, 98
   estate tax, 98
   inheritance tax, 98
debenture stock, 60

Index

debtor nation, 22
defence, 11, 20, 23, 26, 31, 46, 47, 76, 77, 88, 90
de-industrialisation, 65
demand, 28, 34, 35, 52, 70
  aggregate (total), 35, 99
depreciation, 24
devaluation, 99
Diamond Commission, 91-3, 96-7, 106, 107
diseconomies of scale, 64-5
displacement effect, 51
distribution, 15, 16

economic growth, 22, 23, 65, 101
economics, x, 12, 104
  as science, 1-5
  definitions, 1, 10
  laws of, 2
economist, 1, 2, 6, 7, 9, 10, 11, 13, 26, 104
  classical, 34
economies of scale, 38, 58
  defined, 63
  reasons for, 64
economy, 2, 3, 14, 15, 32, 38, 39, 50, 54, 56, 60, 64, 75, 83
  black, 19-20, 94
    see also market, black
  centrally planned, x, 37, 41, 42, 45
  command, 37, 39, 40, 41, 44, 45
    see also centrally planned
  formal, 16-20, 23
  household, 16-20, 23, 24
  informal, 16-17, 19-20, 23, 24
  laissez-faire, 34, 46, 49
  market, 28, 36, 37, 39, 42, 44, 49
  mixed, x, 25, 45, 48, 90, 100, 101, 102, 105
education, 20, 32, 33, 46, 47, 48, 49, 51, 53, 71, 76, 77, 79, 80, 88, 90, 103
  economics of, 4
Elst, P. V., 39, 43, 107
Eltis, W., 65, 107
employment (unemployment), x, 2, 3, 4, 6, 11, 20, 27, 32, 34, 35, 40, 41, 43, 46, 47, 52, 58, 60, 63, 65, 72, 77, 78, 90-3, 98-102
  self-employment, 20, 84, 91, 92
entrepreneur, 27, 28, 29, 30, 32, 35, 36, 37, 57, 58, 60, 88, 95
  origin of term, 106
equilibrium, 27, 34
equities *see* shares
establishment, 63
evasion of tax *see* taxation
exclusion (principle), 31, 32, 47, 48
expectations, 3-4, 28
expenditure, 16, 27
  local government, 75, 79-80
  government *see below* public
  national, 14-25
  public, 49, 50, 51, 54, 55, 75-7, 89, 99, 105-6
  reasons for (public) growth in UK, 51-5
externality (effects), 32, 40, 41, 46, 48, 51, 90

factors of production, 6-8, 12, 17, 26, 27, 28, 29, 34, 37, 38, 42, 47, 57, 64, 93
  see also capital, labour, land, time
Family Income Supplement (FIS), 87
financial services, 65, 67
Financial Statement and Budget Reports, 78
foreign trade *see* international trade
France, 65

Galbraith, J. K., 48, 52, 107
Germany, 44, 65
GDP *see* national product
gift tax, 97
Glushkov, Victor M., 38-9
GNP *see* national product
goods (and services), 4, 15, 16, 17, 21, 23, 25, 26, 31, 32, 34, 37, 46, 47, 48, 57, 78, 93, 99
  consumers', 13-14, 15, 34, 39, 45, 70-1
  private, x, 47-8, 52
  producers', 13-14
  public, x, 47-8, 49, 52, 53, 90
government
  central, 50, 87-8, 90
  local, 50, 54, 75, 79-83, 87-8, 90
government securities (bonds/stock), 3, 15, 85, 95, 105
grants
  to local authorities, 80, 81
  Rate Support Grant, 80
Great Depression, 51, 52

Harbury, C. D., 97, 107
Harrison, A. J., 96, 107
Heath Government, 73
health (services), 33, 46, 48, 49, 51, 53, 76, 77, 88, 90, 103
see also National Health Service
Hicks, J. R., 107
Hitchens, D. M. W. N., 97, 107
housing, 76, 77, 80, 81, 83, 94

income, 3, 5, 6, 14, 15, 19, 20, 21, 22, 27, 28, 35, 36, 52, 78, 92, 95
  distribution of, 38, 39, 40, 46, 51, 90, 91-3
  effect, 93-4
  inequalities in, x, 9, 40, 41, 83, 91-3, 103
  investment, 41, 91, 93, 95
  low, 46, 87
  money and real, 16, 38
  national, 14-25, 50
  policies, 100
income tax, 16, 19, 72, 78, 79, 82, 83, 87, 95
individualist philosophy, 103-4
indivisibility, 64, 67
Industrial Revolution, 58
inflation, x, 4, 11, 52, 54, 90, 98-102
inheritance (inherited property), 35, 41, 97-8
interest, 3, 17, 26, 34-5, 50, 57, 60, 69, 77, 92, 99, 105
international trade, 12, 21-2, 24
investment, 15, 24, 34, 35, 68, 81, 94, 95
  income *see* income, investment
  gross, 14, 25
  net, 14, 24, 25
investment income surcharge, 95
investor, 61, 95, 99
  institutional investor, 61

joint stock company, 58-61
  multinational company, 59-60
  private company, 59
  public company, 59
Jones, G., 82, 107

Kaldor, Lord, 65, 107
Keynes, Lord, 35, 52, 98-9, 107

Labour
  Party, 52, 72, 73-4, 82
  Government, 72, 73, 74
labour, 6-8, 13, 26, 27, 28, 29, 33, 34, 35, 37, 38, 39, 45, 47, 56, 57, 64, 65, 67, 70, 85, 91, 94, 100
  direction of, 42, 47
laissez-faire, x
  see also under economy
land, 6-8, 13, 26, 29, 37, 57, 62
Layfield Report (Report of the Committee of Inquiry into Local Government Finance), 79, 81, 82, 107
limited liability, 58, 59, 60
Lindblom, C. E., 42-3, 107
local authorities/government
  see government, local

manufacturing (industry)/manufacturer, 15, 56, 57, 63, 65, 71
market, x, 26-7, 33, 36, 37, 38, 39, 40, 41, 45, 46, 47, 48, 61, 68, 84
  black, 42
  see also economy, black
  system, 26, 30, 36, 41, 43, 44, 45, 46, 51, 89, 90-1, 101
  see also economy, market
Marshall, Alfred, 1
Marxist (analysis), ix
Meade, J. E., 101, 107
money, 5, 6, 15, 16, 19, 50, 57, 68, 69, 85, 94, 99, 101
  monetarist, 101
  monetary policy, 85, 99, 101
  monetary system, 86
Money Which?, 84, 108
monopoly, 33, 34, 75, 85, 90
moonlighter(ing), 20, 94
multiple stores, 65-7

National and Scottish Bus Company, 83
National Debt, 50, 52, 77, 85
National Economic Development Council (NEDC), 72
National Freight Corporation, 83
National Health Service (NHS), 20, 103
  see also health (services)
National Income and Expenditure Blue Books, 49, 105

112 Index

national product (output), 13–25, 50, 101, 105
  GDP, 65, 83–4
  GNP, 13, 22, 24, 25, 49, 50, 51, 52, 55, 106
  summary of concepts, 25
nationalised industry, 50, 75, 76, 77, 83–5, 103
Niskanen, W. A., 88, 107
Nulter, S. Warren, 38

output, 2, 14, 27, 42
  national output *see* national product
overdraft, 69
owner occupation
  in agriculture, 62
  in housing, 83

partnership, 58–9
Peacock, A. J., 51–2, 53, 107
pension(s), 53, 72, 77, 91
  rights, 95–6
  schemes, 61, 95
Phelps Brown, E. H., 5, 107
Pile, Sir William, 20
pollution, 23, 32, 33, 41, 46, 75, 90
poor relief, 49
population, 22, 53
  working, 70–1
  insured, 99
Post Office, 83–5
poverty, 77, 87, 90
  trap (surtax), 86–7, 90
Prest, A. R., 107
price(s), 2, 4, 6, 23, 25, 26, 29, 30, 32, 33, 39, 41, 43, 49, 52–3, 71, 86, 91, 99, 100, 101, 105
  mechanism/system, 26, 29–31, 33, 91
primary (industry), 15, 56, 61
productivity, 37, 61, 65
  lag, 54
profit(s), 17, 27, 28, 30, 33, 36, 57, 78, 88, 89
Pryke, R., 83–5, 107
Public Accounts Committee, 54
Public Expenditure Survey Committee (PESC), 75–6
Public Expenditure White Papers, 76, 80, 105
public spending
  see expenditure, public

rates (local), 67, 75, 78, 80–3
  domestic rates – arguments for and against, 81
  non-domestic rates – arguments against, 81–2
  rebates, 87
rent, 17, 57, 92
retail distribution/retailer, 56, 57, 66–7
Ricardo, 7
risk, 57–8, 60, 61
Robbins, Lord, 7, 10
Royal Commission in the Distribution of Income and Wealth
  see Diamond Commission

Sandford, C. T., 67, 105, 106, 107, 108
saving(s), 15, 34, 35, 61, 68, 94, 99
scarcity
  definition of, 7–8
  relative, 28, 42
science (scientist), 1, 2, 10, 105
  social, characteristics of, 2–5
Scott, Sir Walter, 58
sector (of the economy)
  manufacturing, 71
  private/market, 54, 55, 59, 61, 68, 98, 102, 103, 106
  public, 47, 49, 50, 51, 54, 55, 56, 65, 71, 75–89, 102
  service, 71
self-employment
  see under employment
shares/shareholders, 15, 59, 60, 95
  ordinary, 60, 61, 95
  preference, 60
Smith, Adam, 30, 35
social balance, 48, 52, 102
socialism (socialist), 44, 52
social science *see under* science
social security, 6, 11, 76, 77, 87, 88
social services, 76, 77, 80
Soviet Union, 38, 39, 43, 45
specialisation, 64
spending *see* expenditure
statistics, 4–5, 14, 16, 18, 23, 24, 25, 61, 92, 105
Stock Exchange, 68, 69
subsidies, 21, 23, 46, 48, 51, 77, 83
substitution, 28, 30, 34, 38, 42
  effect, 93–4

supermarkets, 23, 66-7, 71
Sutherland, A., 61, 108

tax(ation), 3, 19, 20, 21, 25, 46, 47, 50, 51, 53, 54, 75, 78-9, 81, 82, 90, 94, 95, 97, 99
  avoidance of, 94, 97
  evasion of, 19, 94
  see also death duties, income tax, investment income surcharge, rates, VAT, wealth tax
Telecommunications *see* Post Office
Thatcher, Mrs, 54-5, 80
trade cycle, 34
trade union, 34, 56, 70-4, 89, 100, 101, 102
Trades Union Congress (TUC), 72, 73-4
transfer incomes/payments, 20, 21, 46, 47, 50, 92
transfer pricing, 60
Treasury, 54
  Bills, 69, 85

unemployment *see* employment

United Kingdom (UK), x, 15, 18, 20, 22, 25, 46, 47, 48, 49, 52, 53, 56-74, 75-89, 90-102, 105, 106
United States of America (USA), 43, 46, 65
utility, 29, 36, 88

value-added, 15, 57, 65
value-added tax (VAT), 67
value judgement, ix, 9-10, 36, 102-4
Veverka, J., 49, 105, 108
vouchers, 46, 51, 103

wage, 4, 17, 26, 27, 28, 34, 36, 40, 42, 53, 57, 72, 84, 89, 100, 102
wealth
  inequalities in, x, 9, 90, 95-8, 103
  definition of, 95-6
wealth tax, 97
Webb, Sidney and Beatrice, 71
welfare, 9, 22, 23, 42, 72, 77
Wilson, Sir Harold, 70
Wiseman, J., 51-2, 53, 107

Yugoslavia, 46